Palgrave Studies in Impact Finance

Series Editor
Mario La Torre, Department of Management,
Sapienza University of Rome, Rome, Italy

The *Palgrave Studies in Impact Finance* series provides a valuable scientific 'hub' for researchers, professionals and policy makers involved in Impact finance and related topics. It includes studies in the social, political, environmental and ethical impact of finance, exploring all aspects of impact finance and socially responsible investment, including policy issues, financial instruments, markets and clients, standards, regulations and financial management, with a particular focus on impact investments and microfinance.

Titles feature the most recent empirical analysis with a theoretical approach, including up to date and innovative studies that cover issues which impact finance and society globally.

More information about this series at
https://link.springer.com/bookseries/14621

Nadia Linciano · Paola Soccorso ·
Claudia Guagliano
Editors

Information as a Driver of Sustainable Finance

The European Regulatory Framework

palgrave
macmillan

Editors
Nadia Linciano
Commissione Nazionale per le
Società e la Borsa
Rome, Italy

Paola Soccorso
Commissione Nazionale per le
Società e la Borsa
Rome, Italy

Claudia Guagliano
Risk Analysis and Economics
Department
European Securities and Markets
Authority
Paris, France

ISSN 2662-5105 ISSN 2662-5113 (electronic)
Palgrave Studies in Impact Finance
ISBN 978-3-030-93767-6 ISBN 978-3-030-93768-3 (eBook)
https://doi.org/10.1007/978-3-030-93768-3

Cover illustration: Piranka/GettyImages

This Palgrave Macmillan imprint is published by the registered company Springer Nature Switzerland AG
The registered company address is: Gewerbestrasse 11, 6330 Cham, Switzerland

Foreword by Carmine Di Noia

The international commitments made over the last few decades to pursue sustainable development pose significant challenges to all public and private players in the economic and financial system. The public sector is unable to finance the transition to a new model of development and calls on the private sector to channel resources into sustainable activities. For this to happen, however, an information ecosystem for sustainability must be established and operate efficiently. This requires, to name but a few, a shared taxonomy of sustainable activities, salient and standardised disclosure by companies seeking funds to develop sustainable projects or support the green transition of their business, and transparency in the way information intermediaries and third parties certify the sustainability characteristics of economic activities and financial products. However, such an ecosystem cannot develop on its own. The well-known market imperfections and failures justify public intervention.

This book offers a comprehensive and well-argued survey of the public intervention that in the current context is emblematic for its scope and ambition, namely the European Union agenda on sustainable finance. The volume offers a systematic review of the EU holistic approach, which identifies the development of an efficient sustainability information ecosystem as the driver of sustainable finance, capable of mitigating the information asymmetries at the root of bad practices such as greenwashing and allowing the mechanism of trust to consolidate the development of markets for sustainable financial products already underway.

The book also highlights the challenges related to the implementation of the emerging regulatory framework within the European Union, due to the plurality of actors involved, the changes that will be necessary in the preferences and objective functions of the various actors, and the need to map out a gradual path of transition towards sustainable development accounting and possibly mitigating the inevitable redistributive effects that transition will bring within economic and financial systems.

A fundamental role will be played by regulators and supervisors, who must drive market participants towards mainstreaming sustainability into their choices and activities. To this end, it is key fostering a long-term vision capable of taking due account of the interconnections among the various economic and financial players and among the different geopolitical and economic areas of the world.

The book helps to place the EU actions in the context of current market developments, highlighting their global nature just as the challenges posed by the objective of sustainable development are global. It is precisely for this reason that EU action cannot remain isolated. This is another message that the book leaves behind. International cooperation is unavoidable in defining not only sustainability objectives but also the policy actions and tools through which they can be achieved.

Paris, France

Carmine Di Noia[1]
Director of Directorate for Financial and
Enterprise Affairs—DAF
OECD

[1] The information and the opinions contained here pertain to the author and do not reflect necessarily the official position of the OECD.

Foreword by Mario Nava

If in the late Summer 2015, an economist had decided to take a 5-year career break to sail around the world and to come back among us in the late summer 2020, she would be definitely surprised first to discover the term *sustainable finance* and then to see how ubiquitous it is. She will find sustainable finance, and its related concept of ESG principles, ubiquitous not only in economic books but also in all matters of life, when asking for a financial advice or when buying a new car or when renting or buying a house or when shopping at supermarkets. *Per se*, that sustainability has become central to most matters of life is an excellent news. Our lucky sailor economist and most of us would argue that is possibly the best news of the last few years amidst many others not so positive such as biodiversity reduction, climate change disasters, economic recession and of course the health crisis.

This excellent book explains in scientific and yet accessible language, rich in references to the literature, four main features of sustainable finance on which I will also focus this preface. Namely, it explains *why* it happened that sustainable finance became ubiquitous, *who* made it possible, *what* was done and *how* it was done.

Why. The *why* issue is addressed immediately in the first chapters. To all of us, who have been studying classical economics in the first-best world with constrained maximization problems delivering a social Pareto-optimal solution, it is intellectually important to understand immediately why a second-best concept such as sustainable finance has taken over from any type of first-best solution. The authors argue convincingly that there are many variables at play, there is a high numbers of actors, there is a complex intertemporal dimension with several conflicting intergenerational interests and that there are time-dependent drivers of economic activities that a first-best classical solution simply may not exist. While she sailed, the Paris Agreement known as COP21 had identified the objective (keep earth temperature under control) while the subsequent policy decisions have explored some of the means to reach that objective. Policymakers around the world quickly realised that those means could not be found in a first-best world so they settled for means in the second-best world. In other words, the book explains that a competitive market with perfect information is not available to compute accurately "social" prices of, say, greenhouse gas emissions (or any other factor hindering the fulfilment of the sustainable development goals). Therefore, public policy by means of Pigouvian prices and/or taxes will deliver only a sub-optimal result and one needs to accept other ways to nudge economic actors to include ESG considerations in their actions.

Who. The book dedicates several chapters to the EU regulation in and around sustainable finance and to the EU role as leader of the world sustainability campaign. This is very correct. In particular, after the 2015 Paris Agreement, Europe emerged as the champion of sustainable finance and as the one jurisdiction leading the world's efforts towards the fulfilment of the goals agreed in Paris. Europe contributes to only 9% of the world's emission and yet it took on its shoulder the effort to lead the rest of the world to reduce its 91%. Europe took the leadership both for internal and external reasons. Internally, the EU public opinion and the EU leaders are certainly the most attentive and the most demanding in the world to climate change and to social issues. The young students that were marching at Fridays for Future did not only demand Europe to reduce its 9%, but they demanded to act also on the remaining 91%. The EU social market economy is a fertile environment, including in terms of ethical values, to integrate and nourish the sustainability issue and to create a culture of transition towards it. Externally, it appeared clearly that there was no other leader in sight and without a strong EU leadership it would have been difficult, if not impossible, to reach the concrete results that the Paris Agreement intended to reach.

What. Europe lead become stronger as from 2018 with the first concrete EU legislative acts and some external political initiatives directed to the rest of the world.

The launch of the sustainable finance strategy and in particular the very fast adoption of three fundamental pieces of legislations (the EU benchmark regulation, the EU disclosure and financial advice regulation and the EU taxonomy, all very precisely described in this book) gave concreteness to the EU ambitions on sustainability and to the European Green Deal. There are two crucial elements in these legislations, which are transparency and information that are condition-sine-qua-non elements for the result in a second-best environment.

As for the external political initiatives, eight central banks, with the secretariat of the Banque de France and the Chair at De Nederlandsche Bank, launched the initiative of the "Network for Greening the financial system", which today counts 105 members. In parallel, the European Commission set up the International Platform for Sustainable Finance, which counts today 44 authorities representing 55% of the world GDP and 55% of the greenhouse gas emissions of the world. This allowed to scale-up sustainable finance in the global world and to treat it as a global challenge.

How. The EU-mentioned regulations rely on market forces and in particular on information to and from markets and on overall investment transparency. Regulations expose the investors' activities to the markets but do not oblige investors in any way to invest on any activity. Information and transparency improve the second-best solution and alone constitute the powerful incentives to market actors to increase financing of green activities and reduce financing of other activities. It is important to understand why a more coercive approach, as, for example, in the area of capital requirements where Regulation mandates banks' own capital levels and do not only simply expose to the markets banks' own capital levels, could not be adopted. The convincing arguments exposed in the various chapters of this book are essentially twofold. First, the size of the investment needed is so high and so widespread across all world jurisdictions that no single jurisdiction would have neither the economic nor the legal power to mandate them in a coercive manner. Second, an ESG-compatible investment is coherent with profit maximization and therefore using incentives rather than orders is a perfectly efficient strategy. Indeed, because of the inherent long-term interest of investors, full information and transparency on the impact on sustainability of their portfolio are likely to move them towards a more sustainable portfolio.

For that to happen is essential that shareholders and stakeholders in general are aware of the double materiality of information: that is not only what the portfolio does to climate objectives but also what climate change does to the portfolio. If the portfolio is likely to worsen climate change, then investors who follow a purpose will readjust their portfolio. If the future path of climate change is likely to reduce the value of the portfolio, then investors who care for their return will readjust. Double materiality of information speaks to both the investors that care only for the purpose (an altruist) and the investor that cares only for profit (an egoist) and, of course, to any linear combination of these two extremes where all investors can be found. One of the most important message of the book is exactly the complementarity, in the medium to long term, between caring for ESG principles (having a purpose) and realising profits.

Finally, and before leaving the word to the Editors and the Authors, this Preface would not be complete without underlying that this book not only speaks about diversity but also is written and edited by a pool of authors who are collectively a hymn to diversity. There are 14 authors of different backgrounds active in 9 different institutions spread in several EU countries coordinated by three editors who ensured that chapters complete one another to make this book one of the most authoritative compendiums of sustainable finance.

Mario Nava
European Commission[1]
Director General for Structural Reform Support
European Commission
Brussels, Belgium

[1] The information and the opinions contained here pertain to the author and do not reflect necessarily the official position of the European Commission.

Contents

Notes on Contributors

Barbara Alemanni is a Professor of Financial markets and Institutions at the University of Genoa and an Affiliate Professor at SDA Bocconi School of Management. Since May 2020, she is a member of ESMA Securities and Markets Stakeholder Group (SMSG). She is an expert in behavioural economics, securities market micro and microstructure and regulation of financial markets.

Lucia Alessi is responsible for Sustainable Finance at the Joint Research Centre of the European Commission. With her team, she provides scientific evidence in support of EU policymaking in the areas of sustainable investing and ESG risks. Her most recent research focusses on climate stress-testing for financial institutions and the market's attitude towards climate risk, as well as on the assessment of green financial flows and the integration of ESG in banking. From 2007 to 2015, she worked at the European Central Bank, where she served in various DGs, including Research, Economics and Macroprudential Policy and Financial Stability.

Cristiano Busco is Full Professor in Accounting, Reporting and Sustainability at Luiss University in Rome and Luiss Business School. His main research areas are Performance Measurement and Management, Management Control and Reporting, Sustainability Accounting and Impact Evaluation. He is currently a member of the Board and President of MSD Foundation Italy. Cristiano has been the "Inaugural Chair" of the IIRC (International Integrated Reporting Council) academic network. He is currently co-Chair of the Strategy and Integrated Thinking Group of the Value Reporting Foundation, which has recently been established with the merge of the IIRC and SASB—Sustainability Accounting Standards Board. He is the Academic Coordinator of Luiss Observatory GIIM—Governance of Innovation and Impact Management. He is the author of several international publications in the field of Integrated Reporting, Sustainability and Impact. Cristiano's latest report "Purpose to Impact" is expected to be published in early 2022 by the CIMA—Chartered Institute of Management Accountants.

Mavie Cardi is Associate Professor in "Banking and Finance" at the University of Rome "Link Campus"; she holds a Ph.D. in Law and Economics from "Luiss—Guido Carli" University in Rome. She is a member of the Scientific Committee of the Ph.D. course in "New technologies for law, economy and social development", University of Rome "Link Campus".

She is the author of two books and several essays and articles on financial topics. Her main research interests deal with banking regulation, risk management, financial markets and intermediaries, sustainable finance and ESG ratings.

Alessandro D'Eri joined ESMA as a senior policy officer in the Investors and Issuers Department in 2016. In this capacity, he is responsible for ESMA's working groups focussing on supervisory convergence, respectively, for financial and sustainability reporting. Since 2020 he is also co-responsible for the coordination across ESMA of the sustainable finance work stream. Alessandro holds a Ph.D. in business studies from the University of Roma Tre where he graduated summa cum laude in business economics and he is a certified auditor. Prior to joining ESMA, Alessandro worked in the accounting policy department of Generali Group in Trieste and, before then, as technical staff at the International Accounting Standards Board in London.

EscrigOlmedo Elena has a Ph.D. in Sustainability of the Organisations (2013) from the Universitat Jaume I. Currently, EscrigOlmedo Elena works as a Senior Lecturer at the Finance and Accounting Department of the Universitat Jaume I, Spain. Her current research interest is focussed on sustainable finance, SRI and corporate sustainability assessment. She has published scientific papers in high-impact international academic journals and is involved in several externally funded research projects such as Sustainable Market Actors for Responsible Trade (SMART) (2016–2020), funded by the EU's framework programme for research and innovation Horizon-2020. She is also a member of Daughters of Themis and A Member of Sustainability of Organizations and Social Responsibility Management-Financial Markets Research Group.

Giovanna Frati has been working for the Italian Financial Supervisory Authority (CONSOB) since 2002 and she is currently acting as a Deputy Head of the Asset Management Function. She has reached a high level of expertise in the asset management field from a supervision and regulation perspective, with a focus on the fairness of behaviour of Management Companies and on disclosure of funds' offering documents.

She represents CONSOB within European Securities and Markets Authority (ESMA) Investment Management Standing Committee and its Operational Sub-Group and she has actively participated in a number of relevant EU negotiations (inter alia, AIFMD, UCITS and ELTIF review).

She graduated in Business and Economics from Pisa University (magna cum laude) and she completed her background attending courses at Politecnico of Milano, Bocconi University and London School of Economics.

Claudia Guagliano works in the Risk Analysis and Economics Department at ESMA where she leads the Innovation and Products Team in charge of financial and technological innovation, consumer protection analysis and sustainable finance. She previously worked in the international financial relations department of the Italian Treasury and at Consob. She holds a Ph.D. in international economics from Bocconi University, a master in economics from Pompeu Fabra University and a degree in economics from Bocconi University.

Mario La Torre is a Full Professor in "Banking and Finance" and "Sustainable Finance and Impact Banking" at the University of Rome "La Sapienza". His main research areas are banking, sustainable finance and financial innovation. He is also an expert in audiovisual and art financing. He is the editor of the series "Palgrave Studies in Impact Finance". He is currently a member of the Board of the Italian National Body for Microcredit. He has been a member of the G8 Taskforce on Social Impact Investments, a Councellor of the Minister of Culture, a member of the Audiovisual Working Party at the European Council and a member of the Board of Cinecittà Holding. He has also been a member of the consultative group for the definition of the Italian Microcredit Law, and a lawmaker of the Italian Tax Credit Law for the audiovisual industry. He is responsible for the Center for Positive Finance, promoter of the University Alliance for Positive Finance and author of the Blog Good in Finance. He is the author of several international publications in the field of banking, sustainable finance, microfinance and film and art financing.

Nadia Linciano is Head of the Economic Research Department in Consob. She is also CONSOB member of the Italian Financial Education Committee. She graduated from York University (UK) and from the University of Naples Federico II with a D.Phil. in Economics. As for teaching experience, she was an adjunct professor of a variety of topics including corporate finance at LUISS (Rome) and financial markets and institutions at LUMSA (Rome). Her work includes several papers in the areas of economics and regulation of financial markets, published in Italian and international journals. Her research interests include consumer protection; financial education; behavioural finance; financial market regulation; corporate governance; sustainable finance and financial innovation. She participates in ESMA, IOSCO and OECD Task forces working on systemic risks, consumer protection and financial education. She currently coordinates the following CONSOB Reports: Risk Outlook, Report on Corporate Governance of Italian Listed Firms, Report on Non-Financial Disclosure of Italian Listed Firms, Report on Investment Choices of Italian Households and Report on the Financial Advisor-Customer Relation.

Sara Lovisolo is Head of Group ESG at pan-European market infrastructure provider Euronext. She has been an active member of the Consultative Group of the UN-backed Sustainable Stock Exchanges initiative since 2014, and in 2018 joined the Steering Committee of FC4S, the UNEP-supported network of Financial Centers for Sustainability. In June 2018 she was appointed to the EU Technical Expert Group on Sustainable Finance and in 2020 to the EFRAG Task Force tasked with advising the European Commission on the development of an EU non-financial reporting standard. She has a Master's in Economics from Bocconi University and has authored a number of publications on sustainable finance. She also has a post-graduate certificate in Applied Anthropology from the University of Milan-Bicocca.

Julien Mazzacurati is a Senior Economist at ESMA in the Risk Analysis and Economics department. He focusses on sustainable finance analytics, including ESG-related trends and risk monitoring, and participates in several international work streams on climate risks. In addition to risk analysis and research, he supports ESMA's regulatory policy work on sustainable finance. Upon joining ESMA in 2013, Julien worked for several years on European securities, repos and derivatives markets and published several papers on these topics. Before that, he worked for five years at the Institute of International Finance in Washington DC. Julien holds an M.Sc. in Economics from the London School of Economics and a Master's in Management from ESCP Business School.

Valerio Novembre works in Paris as a Senior Policy Officer in the ESMA Investors and Issuers Department and he is an Adjunct Lecturer of Corporate Finance at Sorbonne University. Prior to these assignments, he was an Economist in the Research and Policy Department of CONSOB and served as an Adjunct Lecturer at the European University in Rome and at the University of Florence. Valerio graduated summa cum laude at the University of Florence and subsequently gained an M.Sc. from the London School of Economics and a Ph.D. from the Institutions Markets Technology Institute in Lucca. He has drafted papers on financial regulation for Italian and international journals.

Jacopo Schettini Gherardini, Ph.D. in Corporate Finance, University of Trieste—is the Research Office Director (and Director in Chief) of Standard Ethics Ltd. He started his professional career working for some large banks in London since 1992, covering various international financial markets (both bonds and equities and derivatives) before joining AEI in 2001 as CEO. He started to focus intensively on topics such as corporate governance, sustainable development, ESG ratings and became Directeur Exécutif of Standard Ethics AEI GEIE (Brussels) in 2004. In 2014 he took up his current position at Standard Ethics Ltd. His office currently has the 500 largest listed issuers globally under sustainability ratings. His first methodological publication on ESG ratings dates from 2002 (published in Italy by Sole 24 Ore). His scientific work forms the basis of the Standard Ethics Eating (SER), which was first issued in 2004.

Paola Soccorso is a senior economist at Economic Research Department of the Italian Securities and Exchange Commission (CONSOB). Her main research interests are behavioural finance, households finance, investor education, sustainable and responsible investments and digitalisation of investment services. She is the author of many studies on individual financial choices, based on quantitative, qualitative and experimental analysis. She contributed to the design and analysis of two CONSOB surveys, i.e. the Report on financial investments of Italian households, exploring psychological attitudes, behavioural traits and actual choices of individuals, and the Mirroring survey, examining the relationship between financial advisors and their clients. She has a degree in Political Economics from Bocconi University of Milan; a Master's degree in European Economy and International Finance from the University of Rome Tor Vergata and a previous job experience in banking sector.

List of Figures

List of Tables

List of Boxes

CHAPTER 1

Introduction

Claudia Guagliano, Nadia Linciano, and Paola Soccorso

The international community is increasingly recognising the need to combine economic development with environmental protection and achievement of collective goals. Since 1987, when the report *Our Common Future* was released by the World Commission on Environment and Development, numerous initiatives have been undertaken: the approval of the United Nations 2030 Agenda for Sustainable Development and the Paris Agreement at the Paris Climate Conference (COP21)

C. Guagliano
European Securities and Markets Authority, Paris, France
e-mail: claudia.guagliano@esma.europa.eu

N. Linciano (✉)
Commissione Nazionale per le Società e la Borsa, Rome, Italy
e-mail: linciano@consob.it

P. Soccorso
Commissione Nazionale per le Società e la Borsa, Rome, Italy
e-mail: p.soccorso@consob.it

N. Linciano et al. (eds.), *Information as a Driver of Sustainable Finance*, Palgrave Studies in Impact Finance,
https://doi.org/10.1007/978-3-030-93768-3_1

were cornerstone achievements along the difficult path towards global actions.

At the same time, market participants are progressively incorporating ESG (Environmental, Social and Governance) factors in their decision-making process, aiming at channelling resources to sustainable projects. Still, private resources devoted to sustainability remain insufficient to support the ecological transition in the proportions and at the pace needed to meet the objectives agreed at the global level. Such a financing gap is due to many market failures. Financial systems have been structured within short-term frameworks and horizons while the most pressing environmental and social challenges are long-term. In addition, lack of common terminology, data gaps, scant and non-standardised non-financial information make it difficult to identify and fund sustainable activities.

The European Union (EU) has long acknowledged the role of private finance in channelling resources towards sustainable activities as well as the need for regulatory measures promoting an efficient ecosystem of sustainability-related information. The 2018 *Action Plan on Financing Sustainable Growth* launched many deeply transformative policies, supporting the EU commitment to become the first climate neutral continent in the world by 2050. The *Strategy for financing the transition to a sustainable economy*, adopted in July 2021, follows up on the 2018 initiative and complements the extraordinary public response to the COVID-19 pandemic, the NextGenerationEU programme, stipulating that at least 37% of resources must be allocated to climate and environmental sustainability actions.

Against this backdrop, the present book aims to offer a systematic, updated and critical review of the ambitious EU regulatory interventions. These interventions will be read mainly through the lens of the role that information can play as a driver of transformation. In fact, all market participants need reliable and comparable non-financial information to incorporate sustainability into their decision-making process and to channel resources to sustainable activities correctly priced by financial markets.

The book consists of three parts.

Part I sketches the background conceptual framework underlying finance for sustainability and the role of information as a trigger of sustainable finance. The first point is explored in Chapter 2 that briefly traces the policy and the academic debate about the definition of sustainable development, sustainable finance, sustainable investing and the ESG factors, at the same time highlighting the steady growth of sustainability-linked financial products. The second point is explored in Chapter 3 that delves deeper into the prerequisites for the growth of sustainable finance, i.e. allocative efficiency of financial systems resting on informational efficiency and requiring an efficient ecosystem of information. To this end, Chapter 3 reviews the role of the main categories of agents (from companies to intermediaries, ESG valuation agencies and investors) in the production, dissemination and use of sustainability-related information as well as the evidence and the reasons of market failure in correctly pricing ESG risks factors and, among these, climate change risks. The Chapter also discusses how information efficiency seems to be evolving over time as investors learn about the impact of ESG factors on companies' risks and financial performance and take them into account in their investment choices. This seems to claim the departure from the Efficient market hypothesis (EMH), which has been informing modern finance and investment strategies since its inception, and to advocate for an active investment approach, based on fundamental analysis of companies' social and environmental value (alongside their financial value).

Part II of the book explores the state of play of the production, dissemination and use of sustainability-related information, given the regulatory framework in place before the implementation of the 2018 Action Plan. Chapter 4 focusses on primary non-financial information, as produced and disclosed by companies, and on the main issues that remain unsolved within the current normative environment; in addition, the intertwining between non-financial reporting and corporate governance is explored, particularly with respect to directors' remuneration policies and shareholders' engagement. Chapter 5 is devoted to third parties' information, whose reliability and significance are heavily affected not only by the reliability of primary information but also by the agency's business models and the methodologies used for measuring and evaluating corporate sustainability. The question is therefore what kind of information investors get and how they use the information available to make their choices. Chapter 6 analyses this issue with respect to the category of institutional investors and within the legislative EU framework regulating their decision-making process, whereas Chapter 7 explores the behaviour of retail investors, also in the light of cognitive biases that can impair their ability to access and use both financial and non-financial information.

Part III of the book systematically reviews the current developments in the European regulatory framework aimed to provide an answer to market failures and regulatory issues discussed in Part II. Chapter 8 illustrates the 2018 Action Plan and the 2021 Strategy for financing the transition to a sustainable economy. The EU Action Plan was designed so as to drive a shift in capital allocation towards sustainable assets essentially through the lever of transparency, relying on the assumption that once information on sustainability risks and opportunities has been made available then all financial system actors will inevitably factor in it. Key pillar, among the others, is corporate information, which is discussed in Chapter 9 with respect to the proposal for Corporate Sustainable Finance Disclosure and the other initiatives in place aimed at widening the scope of non-financial reporting obligations and the reliability and comparability of corporate disclosure. Chapter 10 delves deeper into the Action Plan regulatory use-cases or compliance cases for sustainability information, such as the introduction of ESG considerations into investment advice and the requirement for all financial markets participants (broadly, institutional investors) to report on the adverse impacts of their investment and integrate ESG considerations in their governance, investment policy and risk management under the Sustainable Finance Disclosure Regulation.

Overall, the book tries to picture the complexity of the levers that need to be activated to create the proper incentives consistent with sustainable growth. The EU is tackling this complexity through an ambitious and inevitably stratified set of rules, which may turn to be inconsistent across sectors to the extent that they amend existing sectorial legislation, and whose effectiveness claim a global, still to come, legal framework. These remarks are explored in the Conclusions, which also recall the contribution of the new technologies in fostering sustainable finance, and the challenges to both consumer protection (as for the risk of greenwashing and over market bubbles) and market stability (again in terms of bubbles and repricing of stranded assets).

The book lends itself to a thematic key to reading as it includes different paths. Readers interested in corporate governance and corporate disclosure can refer to Chapters 4 and 9. Those willing to deepen sustainable investing can go through Chapters 6 and 7. Economists may appreciate literature review and analysis in Chapters 3, 5, 6 and 7, while lawyers could follow the file rouge guiding through the reference regulatory framework and related policy measures in Chapters 2, 4, 8, 9 and 10. Finally, the law and economics perspective on non-financial information may be enjoyed in Chapters 3 and 5.

PART I

Sustainable Finance: The Conceptual Framework

CHAPTER 2

Sustainable Growth in the European Framework and the Role of Finance

Lucia Alessi, Claudia Guagliano, Nadia Linciano, and Paola Soccorso

Introduction

The notions of sustainable development and sustainable finance find their roots both in the policy debate and in the market practice, initially progressing along separate paths. As soon as finance started to be acknowledged as the key channel of resources towards sustainable activities, efforts on the definition of sustainable investing and on the

Lucia Alessi authored section 'Recent Developments in ESG Markets'.

L. Alessi
European Commission, Bruxelles, Belgium
e-mail: lucia.alessi@ec.europa.eu

C. Guagliano
European Securities and Markets Authority, Paris, France
e-mail: claudia.guagliano@esma.europa.eu

N. Linciano et al. (eds.), *Information as a Driver of Sustainable Finance*, Palgrave Studies in Impact Finance,
https://doi.org/10.1007/978-3-030-93768-3_2

environmental, social and governance factors (ESG) that need to be incorporated into decision-making to promote sustainable development proliferated. However, market participants and investors, in particular, still complain about the lack of homogeneity in definitions and terminology that translates into a lack of comparability in sustainability-related disclosures and investment options. Additional obstacles hindering sustainable finance are the scarcity of information, the underdevelopment of metrics and analytical tools and the short-termism underpinning preferences and choices of economic agents. To drive a shift away towards a new paradigm relying on a long-term vision and channelling resources towards sustainable activities, the European Union (EU) has taken the lead by engaging in an ambitious programme of reforms of the financial markets legislation.

This chapter sketches the evolution of the conceptual framework that underlies the recent advancement of sustainable finance and sustainable regulation in the EU. In particular, section 'Conceptual Framework' will explore the concepts of sustainable development and ESG factors, sustainable finance and sustainable investing. Section 'The EU Holistic Approach to Sustainable Development' will review the EU holistic approach committing, among the others, to an increasingly ambitious objective of reduction of greenhouse gas emissions and recently culminating in the EU Green Deal, intended also as a means for shaping a sustainable recovery from the COVID-19 crisis. Finally, section 'Recent Developments in ESG Markets' provides an overview of recent ESG-related market developments.

Conceptual Framework

Sustainable Development and the ESG Factors

The concept of 'sustainable development' was first defined in 1987 in the Report 'Our Common Future' (also known as Brundtland Report), by the World Commission on Environment and Development (WCED), as

N. Linciano (✉)
Commissione Nazionale per le Società e la Borsa, Rome, Italy
e-mail: n.linciano@consob.it

P. Soccorso
Commissione Nazionale per le Società e la Borsa, Rome, Italy
e-mail: p.soccorso@consob.it

'*development that meets the needs of the present without compromising the ability of future generations to meet their own needs*' (WCED, 1987). In this first sense, sustainable development is linked to the protection of the environment, a principle that will be taken up again at the United Nations Conference on Environment and Development (UNCED), also known as the 'Earth Summit', held in 1992 in Rio de Janeiro, Brazil.

Sustainable development was then consolidated as a principle of international law and contributed to the evolution of international environmental law through the conclusion of regional agreements and global treaties. Among these, the Paris Agreement (COP21) of 2015, adopted by 196 countries, is the first universal and legally binding agreement on a global action plan. Also in 2015, the United Nations adopted the 2030 Agenda, which defines 17 goals (Sustainable Development Goals—SDGs), broken down into 169 Targets, to be achieved by 2030.

As pointed out by several authors (e.g. Beatley & Manning, 1998; Berke & Conroy, 2000; Campbell, 1996), in the first agreements the concept of sustainability remained too vague to allow the development of a normative paradigm capable of guiding policymakers in the implementation of concrete actions. Nor did economic theory helped, since it has long continued to use the notion of sustainability only in diatribe between 'orthodox' economists, on the one hand, and critics of linear economic growth, on the other hand, while referring only to the use of environmental resources (Timpano & Fedeli, 2019). Since the Brundtland Commission, however, there has been a growing consensus among economists that the notion of sustainable development is compatible with the classical economic paradigm. Soppe (2004), for example, believes that it is possible to incorporate the concept of sustainability into the traditional finance paradigm without forcing it if one recognises both 'the storage function of money and capital, which implies that finance [...] is very well suited to realize (or not to realize) "*future generations" needs*' and the financial processes reflecting underlying—medium/long term—real economic processes.

In this way, the classical paradigm can integrate a model of economic growth based on a dynamic and long-term perspective, which takes into account the interests of both current and future generations and recognises the interdependence between the economic, environmental and social dimensions (the so-called 'three Es': 'environment, economy and equity'; Pezzey & Toman, 2005). Over time, there has been a growing awareness that the integration of environmental and social considerations into economic processes cannot disregard the governance of public and private decision-makers, since '*Companies that persist in treating climate change solely as a corporate social responsibility issue, rather than a business problem, will risk the greatest consequences*' (Porter & Reinhardt, 2007). This has led to the recognition of the interdependence between sustainable development and the Environmental, Social and Governance factors (so-called ESG factors), by now universally acknowledged as the pillars of sustainability.

A key issue is what can be considered under the umbrella of the ESG factors. As pointed out by EBA (2021), in spite of the wide usage of the ESG acronym, a single definition is still lacking. As a way of example, an EBA market survey runs in 2019 (EBA, 2020a, 2020b) and the responses received to an EBA consultation conducted between November 2020 and February 2021 show that credit institutions rely on various international frameworks and standards defining ESG factors, while some of them use their own definitions. In particular, EBA (2021, pp. 26–27) counts among the standards that credit institutions refer to 11 international frameworks addressing ESG factors (from the UN SDGs to the Principles for Responsible Investments—PRI—and the International Integrated Reporting Council—IIRC—Integrated Reporting Framework), 6 frameworks specifically addressing environmental factors (including The recommendations of the Financial Stability Board Taskforce on Climate-related Financial Disclosures—TCFD—and the Climate Bond Initiative Climate Bonds Standard) and 10 frameworks specifically addressing social factors (among which the UN Guiding Principles on Business and Human Rights and the eight fundamental Conventions of the International Labour Organization—ILO). There are commonalities among these frameworks that EBA (2021) highlights while warning that depending on their relevance (i.e. materiality) to specific institutions, ESG factors that are not included in the list may still be relevant and as such should be taken into account.

The relevance of these factors and the need to integrate them into the decision-making process of market players stems from their so-called financial materiality, i.e. their potential material impact on the financial performance and/or risk profile of companies (for more details also on the notion of double materiality, see in particular Chapter 8). This leads to the acknowledgement of the interdependence between sustainable development and ESG factors.

The lack of a common definition of the ESG factors has long been an obstacle to the consideration of such issues in market participants' choices (Box 2.1).

Box 2.1: The definition of ESG factors

EBA (2021) identifies five intrinsic features, which may potentially interconnect with each other and characterise ESG factors: the non-financial dimension (this is the case, for example, of the greenhouse gas emissions and equal rights and ethics); uncertainty about their impact in terms of timing, magnitude and persistence of effects; negative economic externalities, not captured in market prices (see Chapter 3); patterns arising from the value chain, i.e. impacts of an entity's activities and its interactions with different stakeholders within its upstream and downstream value chains and increased sensitivity to changes in public policies (so-called transition risk that will be discussed at length in Chapter 3).

Commonalities of ESG factors

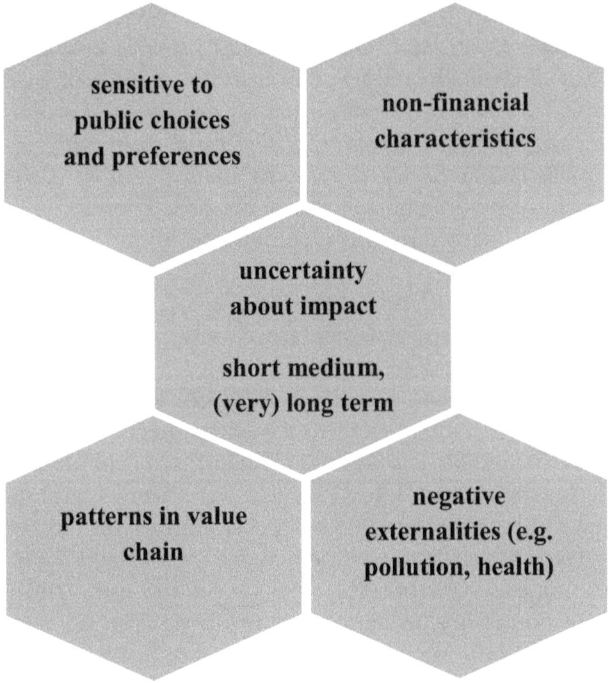

Source EBA (2021, p. 28)

EBA (2021, p. 31) thus defines ESG factors as '*environmental, social or governance matters that may have a positive or negative impact on the financial performance or solvency of an entity, sovereign or individual*'. ESG factors can therefore have negative impacts or positive impacts. Depending on the perspective adopted, either the former (focusing on risks as it is in the prudential regulation) or the latter (focusing on opportunities for financial or non-financial entities related to the transition to a more sustainable economy) can be given emphasis, in line with the need for all market players to take a comprehensive, long-term and strategic approach to ESG issues.

In fact, the fragmentation of definitions and standards concerning the ESG factors give rise to inconsistencies in understanding and management by undertakings that may be particularly marked for the S-factor, '*the most elusive of the triad, being the most general, broad, qualitative, long-term and undefined category of sustainability: in fact, in the absence of a recognised taxonomy, everything that is sustainable but not related to the environment and governance is socially (and generically) sustainable*' (Linciano et al., 2020). So far, developing metrics and KPIs to gauge the S-factor proved to be difficult for several reasons. Therefore, Social issues have usually been characterised on qualitative ground, in contrast to the E-factor.

Difficulties in defining and measuring the S factor are grounded first of all in the fact that social issues are closely related to the principle of corporate social responsibility and thus to a heterogeneous multitude of stakeholders, encompassing workers, consumers, local and global communities. Social claims linked to worker protection have triggered a large number of regulations setting obligations, prohibitions and standards that companies must follow and that have firmly established certain fundamental rights of workers. In turn, this might have temporarily lessened the tension towards the refinement of metrics and KPIs involving other social impacts of a business. In fact, in the early 2000s, Corporate Social Responsibility becomes a top priority in the EU agenda.[1] The aforementioned 2030 Agenda of 2015 also confirmed the role of the social factor among the pillars of sustainability. The following year, the European Commission recalled the 17 UN goals as a reference for all future actions and initiatives.[2] In the context of the European Action Plan to finance sustainable growth, which will be discussed at length in Chapter 8, the European Commission exemplifies ESG factors as follows: '(...) *environmental considerations may refer to climate change mitigation and adaptation, as well as the environment more broadly, such as the preservation of biodiversity, pollution prevention and circular economy. Social considerations may refer to issues of inequality, inclusiveness, labour relations, investment in human capital and communities, as well as human rights issues. The governance of public and private institutions, including management structures, employee relations and executive remuneration, plays a fundamental role in ensuring the inclusion of social and environmental considerations in the decision-making process*'.[3]

The European Action Plan scheduled the analysis of the social issues after tackling the climate emergency. In fact, at the time of writing the EU Platform on Sustainable Finance, set within the framework of the EU Green Deal (see next Section), is sketching a social taxonomy.[4]

The last point concerns the interrelationship among the three factors. In particular, environmental and governance issues have a social dimension for companies. This is clearly acknowledged in the UN 2030 Agenda, pointing out that social issues have both an internal dimension (concerning, for example, management of human resources) and an external dimension to the company (concerning, for example, support to local communities). Both environmental and governance factors may be relevant in the above-mentioned dimensions of the S factor. This implies that pursuing one of the three factors may either facilitate or generate unintended negative consequences on the achievement of the objectives defined with respect to the other two factors.[5]

The Concept of Sustainable Finance

The identification of the concept of sustainable finance has not gone hand in hand with that of sustainable development. The UN's 2030 Agenda, while including a macroeconomic target (Goal 8: Decent Work and Economic Growth), does not explicitly assign relevance to the financial system (Małgorzata et al., 2020), although it does point to investments as a driver of sustainable development.[6]

The G20 Sustainable Finance Study Group, taking its cue from the 2030 Agenda, referred to market mechanisms that foster strong, sustainable and inclusive growth, which can promote the SDGs and improve the stability and efficiency of the financial system through adequate management of risks and market failures (such as those related to the failure in the pricing of negative externalities of business activity; G20, 2016).

The World Bank associates the concept more explicitly with the functioning of businesses and financial markets, drawing on the notion of strong commitment by business owners and managers to integrate sustainability into business strategy (UN Environment and World Bank Group, 2017).

In the European context, as will be explored in the next section, the role of the financial system has been clearly recognised as only private resources can fill the gap to promote the ecological transition that the available public resources are not able to afford. However, the concept of sustainable finance still remains less clearly identified than that of sustainable development. Experts from the High-Level Expert Group on Sustainable Finance indicate in the Final Report (UN Environment and World Bank Group, 2017, p. 6) that: '*Sustainable finance is about two imperatives. The first is to improve the contribution of finance to sustainable and inclusive growth as well as the mitigation of climate change. The second is to strengthen financial stability by incorporating environmental, social and governance (ESG) factors into investment decision-making*'. Furthermore, it is stated that '*Sustainable finance is about a joined-up approach to the development of financial services that integrates the ESG dimensions across market practices, products and policy frameworks*', that is '*axiomatically linked to the long-term*' and it involves '*a commitment to the longer term, as well as patience and trust in the value of investments that need time for their value to materialize*' (UN Environment and World Bank Group, 2017, p. 9).

Following this approach, the European Commission defines sustainable finance as finance that takes into account environmental and social factors in the investment decision-making process, including through an appropriate governance of public and private institutions. In detail, '*Sustainable finance generally refers to the process of taking due account of environmental, social and governance (ESG) considerations when making investment decisions in the financial sector, leading to increased longer-term investments into sustainable economic activities and projects. (...). In the EU's policy context, sustainable finance is understood as finance to support economic growth while reducing pressures on the environment and taking into account social and governance aspects. Sustainable finance also encompasses transparency on risks related to ESG factors that may impact the financial system, and the mitigation of such risks through the appropriate governance of financial and corporate actors*'.[7]

The declination of this approach with regard to all players in the financial system—from companies to investors (both retail and institutional), from financial intermediaries to information intermediaries, from regulators to supervisory authorities—entails an evolution of decision-making processes that embraces both the identification of objectives (towards the creation of financial value in the short and long term) and the reorganisation of production and consumption processes (from linear to circular, i.e. based on production factors including natural and social capital in addition to financial capital). In the more advanced version of sustainable finance, ESG factors are no longer constraints to the maximisation of objectives but objectives themselves (Linciano et al., 2021; Schoenmaker & Schramade, 2019) (Fig. 2.1).

The Concept of Sustainable Investing

Over time, notions referable to sustainability-related investing proliferated leading to many definitions of the so-called sustainable and responsible investing (SRI). SRI is broadly considered to be an investment style combining financial goals with social and environmental objectives (Sparkes, 2001). ESG investing, on the other hand, has a singularly financial motivation (see also Box 6.1 in Chapter 6).

Eurosif blends these approaches by characterising SRI as '*a long-term oriented investment approach which integrates ESG factors in the research, analysis and selection process of securities within an investment portfolio. It combines fundamental analysis and engagement with an evaluation of ESG factors in order to better capture long term returns for investors, and to benefit society by influencing the behaviour of companies*' (Eurosif, 2018). This definition, therefore, refers to a long-term approach that aims both at financial returns for the investor and at creating value for society as a whole, through a commitment of orienting issuers towards the consideration of ESG factors.

The Eurosif (2018) classification, which is similar to definitions of other international networks and associations (e.g. the Global Sustainable Investment Alliance—GSIA, the Principles for Responsible Investment—PRI and the European Fund and Asset Management Association—EFAMA), encompasses several investment strategies (e.g. ESG investing, impact investing and so on) attaching importance not only to the financial dimensions of the investment but also to the use of the funds raised by the

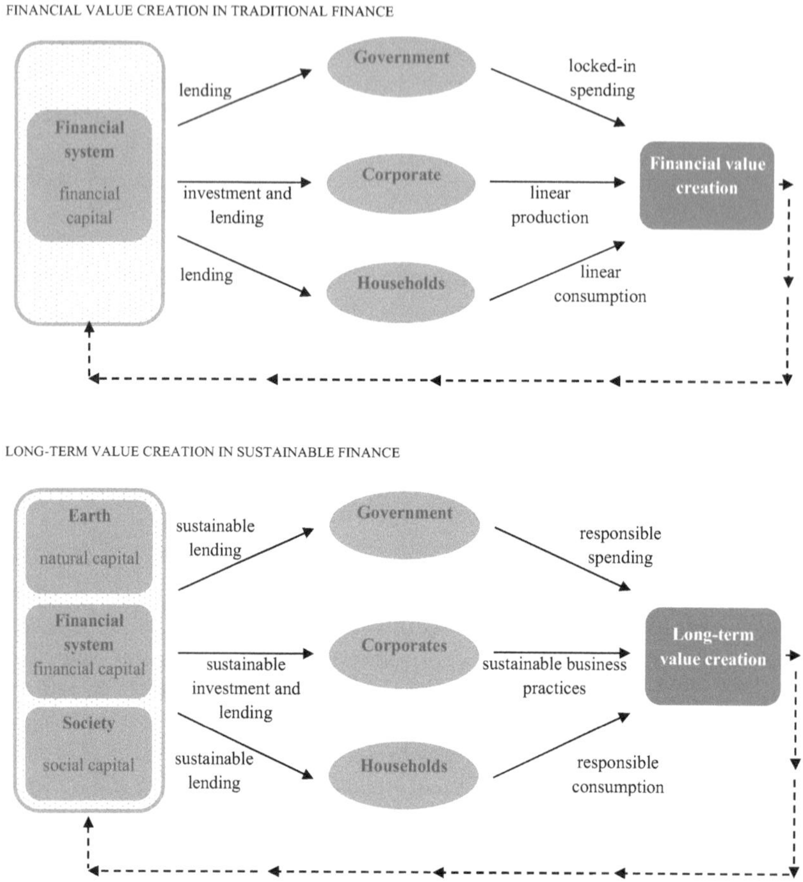

Fig. 2.1 From traditional finance to sustainable finance (*Source* Schoenmaker & Schramade, 2019)

issuers and their corporate governance practices (from positive/negative screening to ESG integration) (see Chapter 6).

In recent years, sustainable investments have been increasing progressively,[8] although many issues may hinder their development. As it will be discussed further in Chapter 8 and over this volume, the many initiatives by the European legislator are intended to remove such hurdles and make viable the transition to a green, low-carbon economy and, in general, to a sustainable development model.

The EU Holistic Approach to Sustainable Development

The European Union has long been concerned with sustainable development and in particular with climate changes. The EU Climate Law[9] marked the intention to become a global leader in the transition towards a net zero greenhouse gas emissions economy. After the peak in the EU greenhouse gas emissions in 1979, several actions were taken to enhance energy efficiency and promote fuel switch and renewables. In addition, already in 2009, the EU decided to curb its emissions by 80–95% by 2050. These policies resulted, between 1990 and 2016, in a cut in energy use by almost 2% and by 22% in greenhouse gas emissions. As of 2018, the EU is responsible for 10% of global greenhouse gas emissions.[10]

In November 2018, the Commission set out its vision for a climate-neutral EU, covering nearly all EU policies and aiming at keeping the global temperature increase to well below 2 °C and pursuing efforts to keep it to 1.5 °C, in line with the Paris Agreement.[11] This long-term strategy confirmed Europe's intention to lead in global climate action by achieving net zero greenhouse gas emissions by 2050.

Policy Actions in the Framework of the EU Green Deal

The EU commitment to become the first climate-neutral continent in the world by 2050 underpins the European Green Deal, an action plan aiming at making the EU's economy sustainable.[12] It designs a set of deeply transformative policies that will be needed to reach this objective, starting with increasing the EU's climate ambition for 2030 and 2050. However, the scope of the EU Green Deal is much broader than climate alone, embracing the environmental dimension as a whole. Indeed, the main elements of the action plan, over and above the 2030 and 2050 emission objectives, cover the following: supplying clean, affordable and secure energy; mobilising industry for a clean and circular economy; building and renovating in an energy- and resource-efficient way; accelerating the shift to sustainable and smart mobility; designing a fair, healthy and environmentally friendly food system; preserving and restoring ecosystems and biodiversity and creating a toxic-free environment.

The COVID-19 pandemic strengthened further this commitment as one-third of the 1.8 trillion euro investments from the NextGenerationEU Recovery Plan, and the EU's seven-year budget will finance the European Green Deal (Fig. 2.2).

The centrepiece of the European Green Deal is the already mentioned EU Climate Law.[13] With the adoption of the Climate Law, the goal set out in the European Green Deal Communication—for Europe to become the first climate-neutral continent by 2050 becomes a legal commitment. The Climate Law also increases the EU's GHG emission reduction target for 2030 to at least 55% compared with 1990 levels, from the previous 40%, and tasks the Commission with an assessment of how the EU legislation would need to be amended in order to reach the 2030 and 2050 targets. Enshrining the 2050 climate neutrality objective in EU legislation ensures that all EU policies contribute to this goal, while taking into account the importance of promoting fairness and solidarity among Member States, and that the transition to net zero is irreversible. A legally binding target, which is set for the EU as a whole, will also bind Member States to take the necessary measures. The Climate Law proposal empowers the Commission to set out a trajectory at the EU level to achieve climate neutrality by 2050, which will be reviewed regularly. It also addresses climate change adaptation, with a view to strengthening resilience and reducing vulnerability to more frequent and severe climate-related natural disasters, which are indeed the objectives of the new EU Adaptation Strategy.[14] Finally, the Climate Law also includes the creation of a system for monitoring progress and to take further action if needed, which is based on existing systems such as the governance process for Member States' national energy and climate plans, regular reports by the European Environment Agency, and the latest scientific evidence on climate change and its impacts.

In July 2021 the Commission adopted a package of proposals, known as the 'Fit for 55' package, to make the EU's climate, energy, land use, transport and taxation policies fit for reducing net greenhouse gas emissions by at least 55% by 2030. Several proposals concern the EU Emission Trading System (ETS), including a decrease of the overall emission cap, the creation of an ETS for road transport and buildings and the introduction of an obligation for Member States to spend their emissions trading revenues on climate and energy-related projects.[15] The package also includes proposed amendments to the Effort Sharing Regulation, assigning strengthened emissions reduction targets to each

Fig. 2.2 Policy actions in the framework of the EU Green Deal

Member State,[16] and to the Regulation on Land Use, Forestry and Agriculture, setting an overall EU target for carbon removals by natural sinks,[17] while the new EU Forest Strategy sets out a plan to plant three billion trees across Europe by 2030.[18] With respect to the energy sector, the proposed amendments concern the following: (i) the Renewable Energy Directive, with an increased target to produce 40% of EU energy from renewable sources by 2030 and strengthened sustainability criteria for bioenergy,[19] (ii) the Energy Efficiency Directive, with a more ambitious annual target for reducing energy use at EU level and an obligation for the public sector to renovate 3% of its buildings each year (see also below on the Renovation Wave Strategy),[20] and (iii) the Energy Taxation Directive, with the removal of fiscal incentives that currently encourage the use of fossil fuels.[21] Looking at the transport sector, the Fit for 55 package proposes to strengthen CO_2 emission standards for cars and vans, de facto banning the sale of internal combustion engine vehicles as of 2035,[22] and to revise the Alternative Fuels Infrastructure Regulation by requiring Member States to expand charging capacity.[23] Two proposed regulations, namely the ReFuelEU Aviation Initiative[24] and the FuelEU Maritime Initiative,[25] will stimulate the uptake of sustainable jet and maritime fuels. Finally, the Fit for 55 package includes a proposed Regulation establishing a Carbon Border Adjustment Mechanism, putting a carbon price on imports of targeted products.[26]

The supply of clean energy will be critical to reach the 2030 and 2050 climate objectives. To foster the clean energy transition, the Commission has launched three initiatives. The first is the EU Strategy for Energy System Integration, which will make the European energy system more efficient and circular, by reducing energy waste, as well as cleaner, thanks to renewable energy production and renewable and low-carbon fuels, such as hydrogen.[27] The strategy notably includes a possible proposal for the extension of the Emission Trading System to new sectors (as already mentioned), as well as the review of the State aid framework and the revision of the Energy Taxation Directive, with a view to phasing out direct fossil fuel subsidies. The second initiative is the Hydrogen Strategy, which will promote the use of hydrogen by developing an investment agenda, boosting demand and scaling up production; designing suitable support schemes, market rules and infrastructure and promoting research and innovation.[28] The third initiative, called European Clean Hydrogen Alliance, also aims at supporting the development of a hydrogen industry in Europe, by pooling resources for scale and impact and creating a pipeline of viable large-scale investment projects.

With respect to buildings, which account for a large share of energy consumed, the Commission launched a Renovation Wave Strategy to improve the energy performance of the building stock.[29] In particular, the strategy aims to double annual energy renovation rates in the next ten years. In parallel to the strategy, the Commission adopted new rules for the smart readiness of buildings, to promote digitally friendly renovations, integrate renewable energy and enable measurements of actual energy consumption.[30] Finally, it launched the New European Bauhaus initiative, an environmental, economic and cultural project, aiming to combine design, sustainability, accessibility, affordability and investment.[31]

With respect to transport, the objective of the EU Green Deal is a 90% reduction in GHG emissions by 2050. To this aim, the uptake of clean vehicles and alternative fuels needs to be boosted considerably and transport needs to shift to more sustainable modes such as rail and inland waterways. The Connecting Europe Facility, the EU funding instrument for strategic investment in transport, energy and digital infrastructure, supports the transition to sustainable mobility by spending 60% of the budget on infrastructure projects with a link to sustainability. With respect to air transport, the Single European Sky initiative is expected to significantly reduce aviation emissions by reducing flight delays and increasing flight efficiency.

Turning to EU Green Deal objectives beyond climate, with respect to the objective of halting biodiversity loss, the Commission has launched the EU Biodiversity Strategy.[32] Among the key commitments of the strategy, there is the establishment of protected areas corresponding to at least 30% of land and sea in Europe, with legally binding nature-restoration targets providing stricter protection of EU forests. Restoration of degraded ecosystems at land and sea will be pursued by increasing organic farming and biodiversity-rich landscape features on agricultural land, halting and reversing the decline of pollinators, reducing the use and harmfulness of pesticides by 50% by 2030, restoring at least 25,000 km of EU rivers to a free-flowing state, and planting 3 billion trees by 2030. Overall, €20 billion per year will be unlocked for biodiversity through various sources, including EU funds and national and private funding.

With the objective of ensuring a healthier and more sustainable EU food system, the Commission has launched the Farm to Fork Strategy.[33] The EU's goals are to reduce the environmental and climate footprint of the EU food system and strengthen its resilience, ensure food security in the face of climate change and biodiversity loss and lead a global green transition. To ensure that these objectives are reached, the strategy sets clear 2030 targets. With respect to pesticides, it aims at reducing the use and risk of chemical and more hazardous pesticides by 50%. With respect to the excess of nutrients in the environment, which is a major source of air, soil and water pollution, the strategy targets are a reduction of nutrient losses by at least 50% and a reduction of fertiliser use by at least 20%. The strategy also aims at reducing the sale of antimicrobials for farmed animals and in aquaculture by 50%, and to increase farmland being used for organic farming to 25% of the total.

Sustainable agriculture is another key policy area covered by the EU Green Deal. Sustainability in agriculture and rural areas across the EU will be ensured through the Common Agricultural Policy (CAP). In particular, 40% of the CAP budget will be climate-relevant, with new funding and additional incentives for climate- and environment-friendly farming practices being unlocked through targeted eco-schemes, and an enhanced conditionality linking EU-funded income support to environment- and climate-friendly farming practices and standards. With respect to rural development, EU support will aim to enhance ecosystems and promote resource efficiency. National CAP strategic plans will combine a wide range of local and EU-level objectives and the Commission will ensure the plans reflect EU environmental and climate-related legislation and address the key quantified European Green Deal targets.

Aiming at cutting air, water and soil pollution, the Commission will adopt the Zero Pollution Action Plan. Based on the roadmap put forward for consultation, the plan will include measures to strengthen implementation and enforcement of relevant EU rules, improve the existing legislation related to health and environment, strengthen the monitoring and governance of pollution prevention and reduction policies, and explore means to drive a societal shift towards sustainable solutions.

The European Green Deal also acknowledges that sustainability will need to be mainstreamed in all EU policies. For example, Horizon Europe will have a central role in leveraging national public and private investments, with 35% of its budget supporting the research and innovation efforts needed to develop new climate-related technologies and solutions. With respect to Member States' public finances, the European Green Deal also promotes a larger use of green budgeting tools and creates the context for broad-based tax reforms, including removing subsidies for fossil fuels, shifting the tax burden from labour to pollution, and taking into account social considerations. The action plan also opens a discussion on how green investments could possibly be treated within EU fiscal rules, while preserving debt sustainability. Looking at the international dimension, Europe is at the forefront of efforts on mainstreaming sustainability at the global level. To this aim, the International Platform on Sustainable Finance was founded, which now counts 17 member countries representing 55% of GHG emissions, 50% of the world population and 55% of global GDP.

Various policy tools will support the financing of the European Green Deal. The European Green Deal Investment Plan, also known as Sustainable Europe Investment Plan, is the investment pillar of the Green Deal.[34] The plan aims at mobilising at least €1 trillion in sustainable investments over the next decade, with half of the financing coming directly from the EU budget, and half being triggered by the EU budget. In particular, the InvestEU fund will dedicate at least 30% of mobilised investments to climate- and environment-related projects, and provide an EU budget guarantee to partially cover the risk of financing and investment operations. Moreover, the European Investment Bank—the Union's climate bank—will gradually increase the share of its financing dedicated to climate action and environmental sustainability to reach 50% of its operations in 2025. The Just Transition Mechanism is the part of the European Green Deal Investment Plan targeted to ensure a fair and just green transition. It will mobilise at least €100 billion in investments over the period 2021–2027 to support the regions most impacted by the transition.

Finally, private investment will be mobilised thanks to several policy actions aimed at promoting sustainable finance, as well as the launch of a Renewed Sustainable Finance Strategy (see Chapter 8).

A Sustainable Recovery

Just after the EU Green Deal was announced the COVID-19 crisis broke out, clearly showing how critical it is to strengthen the sustainability and resilience of our society and economy. Indeed, similar health emergencies will become more and more likely in a context characterised by environmental degradation, biodiversity loss and climate change impacts. Against this background, the European Union made clear that the recovery will be a sustainable recovery because decisive action to fight climate change cannot be postponed further. The response to the COVID-19 crisis is in fact also an opportunity to 'bounce forward' to a sustainable and more resilient development paradigm. Moreover, the response to the COVID-19 crisis will involve heavy borrowing on the capital markets to support the economy in the face of the deepest recession since World War Two. Society cannot afford that these money finances assets that that are doomed to become stranded—all the more so considering that this debt will be paid back by the next generation.

To deliver on a sustainable recovery, the regulation establishing the Recovery and Resilience Facility—the centrepiece of the €800 billion recovery package called NextGenerationEU—asks Member States to design their Recovery and Resilience Plans by ensuring that at least 37% of the funds support measures that address climate change. In addition, no measure should lead to significant harm to environmental objectives as defined in the Taxonomy Regulation (see Chapter 8).[35] Further guidance clarifies that Member States need to provide a 'do no significant harm' assessment covering the direct and primary indirect impacts of each and every measure in their plan, including both reforms and investments. In doing so, they can, but are not obliged to, refer to the technical screening criteria as set out in the relevant Delegated Act under the Taxonomy Regulation.[36]

Building on this obligation for Member States, the Commission has announced the NextGenerationEU Green Bond framework.[37] Under this framework, which will be aligned to the EU Green Bond Standard as much as possible, the Commission will fund 30% of the NextGenerationEU programme by issuing green bonds. This could amount to up to €250 billion of green bond issuance, which will make the Commission one of the largest green bond issuers globally. The Commission's issuance will also significantly boost the green bond market globally and drastically increase the availability of a safe and sizeable green asset, enabling portfolio managers to further diversify their green portfolios. The proceeds raised through the issuance of green bonds will be used to finance the measures in the Member States' Recovery and Resilience Plans which are identified as supporting climate objectives. To ensure that the money is used for green investments, Member States will be required to provide relevant information twice a year.

In parallel, the Commission is already issuing social bonds to finance the Support to mitigate Unemployment Risks in an Emergency (SURE) programme, which aims at providing financial assistance up to €100 billion to help Member States protecting jobs in the face of the pandemic. The four SURE bond issuances so far, totalling an amount of nearly €40 billion, were all largely oversubscribed, resulting in very favourable pricing terms and negative yields for shorter maturities.

Another key lever to mobilise financing for the Green Deal and shape a sustainable recovery is the European Commission Renewed Sustainable Finance Strategy, aimed at providing the policy tools needed for the transition towards sustainability, in a context of recovery from the negative social and economic consequences of the pandemic (see Chapter 8 for further discussion).

Recent Developments in ESG Markets

Over time, sustainable investing has been evolving along two directions. On the one hand, there are bonds, which assign the investor a pure funder role and which see in green bonds as the most widespread category.[38] On the other hand, there is the equity investment, which may result from the adoption of different investment strategies (see below) aimed at picking companies whose business model is somehow consistent with sustainability; equity investment gives the investor the role of a shareholder, who can influence corporate choices also through active engagement (Box 2.2).

Box 2.2: Sustainable products for sustainable investing

OECD (2021) reviews 39 transition finance-related instruments and products, including also instruments and products with an explicit 'transition' label, as well as those positioned as providing transition financing. Among such instruments, sustainability-related bonds are debt instruments complying with the global certification standard operated by the global consortium of the Climate Bond Initiative (CBI), a branch of the International Capital Market Association (ICMA). The ICMA certification is granted to eligible projects aligned with four core components concerning the use of proceeds, processes for evaluating and selecting projects, management and subsequent reporting to the public.

This approach applies to green, social and sustainability bonds. Green bonds are instruments whose proceeds or an equivalent amount are exclusively used to finance or re-finance, in part or in full, new and/or existing eligible Green Projects. Social bonds, which gained momentum during the pandemic crisis, are designed to finance employment protection schemes (e.g. the EU SURE—Support to Mitigate Unemployment Risks in an Emergency). Sustainability bonds included in the spectrum of the financed activities projects are aimed at environmental and social goals.

A further type of bonds is represented by sustainability-linked bonds, with financial and/or structural characteristics varying depending on whether the issuer achieves predefined Sustainability/ESG objectives within a predefined timeline (ICMA defines them as a forward-looking performance-based instrument). For such bonds, the core components refer to the selection of Key Performance Indicators (KPIs), the calibration of sustainability performance targets, bond characteristics, reporting and verification.

Finally, green and sustainability-linked loans are loans tailored to the needs of the companies gathering funds.

Looking at sustainable investing, the universe of ESG funds includes both ESG labelled funds and non-labelled funds following ESG strategies. This covers funds that consider companies' ESG in their investment policy (through positive or negative screening), funds actively engaging with companies on sustainability-related matters, impact funds with predefined ESG metrics and targets, and specialised funds investing only in, e.g. renewable energies or clean water assets.

Within labelled funds, a high degree of heterogeneity exists across EU countries due to the absence of an EU-wide ESG label. ESG fund labels exist for example in Austria, France, Germany and Luxembourg but rely on different criteria. In some EU countries, ESG labels do not exist. These country-specific situations raise consistency issues when funds are marketed beyond national borders.

For non-labelled funds, data providers rely primarily on fund documentation (prospectus, Key Investor Information Document KIID, etc.). However, this does not guarantee that the funds achieve their sustainability objectives nor does it allow for the benchmarking of sustainability performance against objective criteria. This could result in misalignment between investors' sustainability goals and their investments.

As for green bonds, since 2014 till 2020, yearly issuances at a global level grew from EUR billion 27 to EUR billion 257, more than doubling over June 2020 to June 2021. In a first stage, the most active issuers have been supranational financial institutions, such as the World Bank and the European Investment Bank, while in recent times, both sovereign issuers (OECD, 2020) and private companies have started to contribute significantly, as evidenced also by the increase in the number of companies issuing multiple green bonds (ESMA, 2021).

Europe dominates the green bond market with issuances accounting for around 60% of the global aggregate (as of June 2021), also reflecting the European Commission's decision to finance the EU decarbonisation programmes through green bonds (Fig. 2.3).[39]

As for sustainable equity investing, the phenomenon is captured by the so-called ESG funds. Such funds have effectively gone from a niche investment idea to the mainstream. Sustainable investing is also increasingly accessible, through a broadening array of investment vehicles, including those suitable for retail investors. The growth of sustainable investment can be attributed to a combination of factors. First, investors have experienced an increased level of interest in ESG topics from their underlying clients over the last couple of years and the expectation on investment decisions to comply with certain standards has grown, also due to greater government and regulatory pressure, most notably stewardship codes.

Second, there is an increasing belief that some ESG issues can be important indicators of investment risk and the ability of a company to create value over the long term. Amel-Zadeh and Serafeim (2018) highlight that relevance to investment performance is the most frequent motivation for use of ESG data followed by client demand and product strategy, bringing change in companies and then ethical considerations. Furthermore, they point to differences across the various forms of sustainable investing, with ESG integration being driven by relevance to investment performance, when screening is mostly driven by product and ethical considerations. State Street Global Advisors (2017) drawing from a survey of close to 500 institutional investors demonstrates that ESG factors are regarded as financially material by most investors. In its 2019 letter to CEOs, Larry Fink stated '*Profits are in no way inconsistent with purpose—in fact, profits and purpose are inextricably linked. Companies that fulfill their purpose and responsibilities to stakeholders reap rewards over the long-term*'.

The growth of ESG funds was facilitated also by the development of sustainability or ESG indices or benchmarks, which differ from traditional

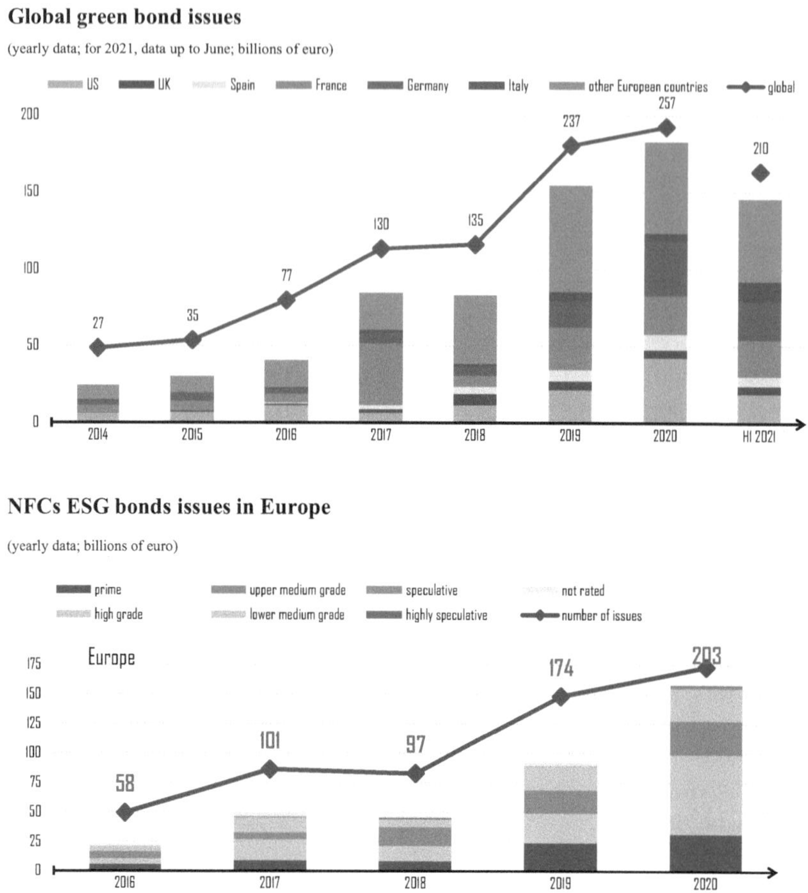

Fig. 2.3 Global green bond issues (*Source* Refinitiv Eikon. Dealogic. Rating of issues: 'prime' = AAA; 'high grade' = AA−/AA/AA+; 'upper medium grade' = A−/A/A+; 'lower medium grade' = BBB−/BBB/BBB+; 'speculative' = BB−/BB/BB+ (non-investment grade); 'highly speculative' = B−/B/B+)

broad market indexes as they include ESG criteria in security selection. Such indices and benchmarks made also viable the development of passively managed ESG funds (Box 2.3).

Box 2.3: ESG indices and benchmarks

The development of sustainable investing has been accompanied by the spreading of ESG indices and benchmarks, covering most global stock exchanges and are now also available at industrial and cross-sectoral levels. The objective of such indices and benchmarks is to provide a transparent and rules-based way to build an investment strategy with ESG characteristics, e.g. avoiding certain kinds of companies or gaining exposure to high ESG ratings.

There are a huge variety of indices available on the market: a small sampling is reported in European Commission (2020) and includes Bloomberg Gender Equality Index; FTSE ESG Index Series; Solactive ISS ESG EU, Climate Benchmark Index Series, Solactive ISS Low Carbon Index Series, Solactive ISS.

ESG Screened Index Series; MSCI ESG Leaders Index, MSCI ESG Focus Index, MSCI.

Climate Change Index; Refinitiv/S-Network ESG Best Practices Indices etc. Within the EU, investable ESG indexes are mostly pan-European, such as the STOXX Europe ESG Leaders 50 Indices or the Dow Jones Sustainability Europe Index.

Financial instruments included in the indices are identified on the basis of the ESG performance of their issuers. Among the criteria used most frequently, there are the exclusion of issuers or entire sectors characterised by core not in line with sustainability issues (e.g. arms and fossil fuels sectors) or openly not in line with sustainability issues (e.g. arms and fossil fuels sectors); the concentration on one or more sustainability issues (e.g. climate change and energy efficiency); the identification of the best-in-class issuers and the inclusion of companies achieving a certain ESG rating level.

The methodology can be either backward or forward-looking. In the first and more common case, evaluation factors are based on past performance. In the second case, evaluation factors are based on expected future performance in a given scenario.

The ESG benchmarks thus constructed are typically divided into equities, bonds and thematic benchmarks (a subset of equity benchmarks). Beyond allowing the design of products characterised by financial instruments that are consistent with sustainability features (mutual funds, ETFs, etc.), these indices make it possible the comparison with traditional indices in order to verify the effect of the integration of ESG criteria into the investment process.

During the Covid crisis in the first half of 2020, there have been several interesting developments in ESG equity markets. ESG equity indices have shown resilience during the crisis. As of end-June 2020, gross annual returns for the MSCI World ESG Leaders index and Euro Stoxx ESG Leaders 50 were 350 bps and 450 bps, respectively, above their non-ESG peers (ESMA, 2020).

This resilience can be explained to an extent by sectoral developments. Fossil-fuel producers have suffered from the oil price crash while the lockdown had an outsized impact on energy-dependent sectors such as transportation. Meanwhile, technology and healthcare companies saw demand for their products soar, while they tend to be overrepresented in ESG indices and portfolios. However, this does not entirely explain the differential or the consistent outperformance of ESG indices over the last few years. Other explanations put forward by market observers include the capacity of ESG leaders to innovate, prudent risk management, long-term targets and flexible internal policies (e.g. teleworking), making these companies generally better equipped for the new situation (ESMA, 2020).

As ESG investing continues to take hold, a wide array of new products is becoming available in the market. Offerings of EU ESG ETFs, in particular, jumped in 2020 to 90, exceeding for the first time the number of new non-ESG ETFs (62) as some ETF providers delayed the launch due to uncertain market conditions. Although ESG ETFs only hold 12% of EU ETF assets, they are catching up fast with an annual growth rate above 200% in 1H21 versus 27% for non-ESG ETFs (ESMA, 2020; see Fig. 2.4).

The growth in ESG fund assets over the last two years also highlights the declining popularity of exclusions-based only strategies (+5% AuM in 2020)—whereby funds exclude entire sectors or companies (e.g. due to their involvement in controversial or unethical activities; more on the investing strategies in Chapter 6). Instead, a growing share of funds now combine such exclusions with ESG fund strategies (+27% AuM in 2020; ESMA, 2021). Among ESG strategies, impact funds (aiming to achieve quantitative ESG targets) and funds focused on environmental sectors have experienced particularly strong growth, with AuM up 37 and 48%, respectively, for these strategies in 2020, which now represent around 40% of ESG fund assets. Disclosure requirements under SFDR bring further transparency to the implementation of ESG fund strategies and their impact on environmental and social aspects (see also Chapter 7).

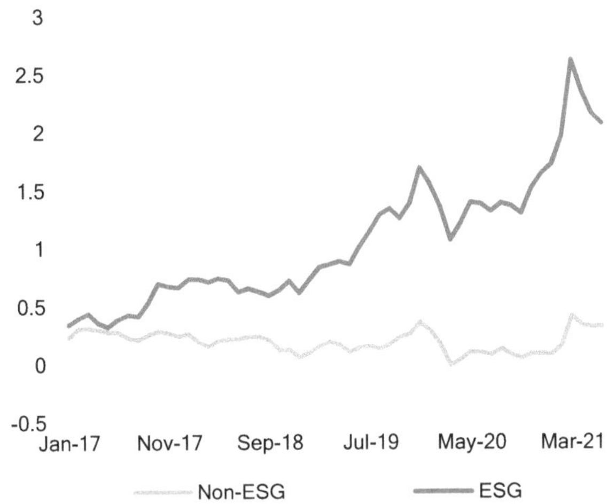

Fig. 2.4 Chart ESG ETF (*Note* Annual AuM growth of EU-domiciled ESG and non-ESG ETF, in %. *Source* Morningstar, ESMA)

Conclusions

In recent years, sustainable investing has been experiencing a steady growth, amid a market-driven ecosystem, with an increasing number of firms engaged in corporate and social responsibility (CSR), an increasing number of institutions providing ESG data, benchmarks and ESG rating and scores, and some initiatives aimed at providing global standards (especially at the product level). However, the lack of a shared taxonomy and KPIs identifying sustainable project and activities of agreed reporting standards both at the company and the product level and of a common regulatory framework have raised many obstacles to the development of sustainable finance.

There has been an increase in the supply of funds and bonds whose ESG label is the result of often inconsistent assessments: IMF (2019) shows that only 37% of funds classifiable as ethical according to Lipper are also sustainable according to Bloomberg. Similar evidence emerges for bonds linked to the financing of sustainable assets. Moreover, in recent times, there has been a proliferation of ESG bond types, identified by as many labels, without this corresponding to the establishment of a homogeneous classification shared by market participants. Among the most widespread standards are those of the already mentioned Climate Bonds Initiative, one of the largest international not-for-profit organisations that collect and manage data on green bonds globally. These circumstances make it difficult to compare multiple products that qualify as sustainable and increase reputational and greenwashing risks.

Finally, the difficulty of measuring the ESG impact of portfolio management, given the aforementioned proliferation and fragmentation of benchmarks offered by market operators, is a further critical issue that the European legislator has taken into account (along with others) in the Action Plan for Sustainable Finance, as will be seen at length in Chapter 8.

Notes

1. This occurred first with the Lisbon European Council in 2000 and then with the Green Paper Promoting a European framework for Corporate Social Responsibility in 2001, which announced the goal of sustainable, social and economic development. In that occasion, it was stated that in addition to creating value for shareholders, the company must preserve environmental, social and human capital over time by optimising resources.
2. See Communication from the Commission to the European Parliament, the Council, the European Economic and Social Committee and the Committee of the Regions, *Next steps for a sustainable European future. European action for sustainability.* Strasbourg, 22.11.2016, COM(2016) 739 final. Retrieved from https://ec.europa.eu/eurostat/documents/8105938/8292490/communication-next-steps-sustainable-europe-20161122_en.pdf/427f429a-5bc7-4476-a684-09da17ebc24f.
3. See https://ec.europa.eu/info/business-economy-euro/banking-and-finance/sustainable-finance/overview-sustainable-finance_it; accessed on 28 February 2021.
4. See the EC Platform on Sustainable Finance, Draft Report by Subgroup 4: Social Taxonomy, July 2021, retrieved from https://ec.europa.eu/info/sites/default/files/business_economy_euro/banking_and_finance/documents/sf-draft-report-social-taxonomy-july2021_en.pdf. For more details, see Chapter 8.
5. In fact trade-offs between the different SDGs may exist. For example, the transition to a green economy and the consequent decarbonisation implies a strong containment of the extraction and use of fossil fuels, which in turn generates income and wealth distributive effects thus impairing poverty eradication in the involved areas (Linciano et al., 2021).
6. Following the institutional debate that will be sketched in the following, sustainable finance may be defined in its broadest sense as finance

enhancing the financial stability of an economy by integrating ESG factors and a long-term view into decision-making processes. In a narrower sense, it is associated with the reallocation of resources in favour of low-carbon investments (Małgorzata et al., 2020).

7. See https://ec.europa.eu/info/business-economy-euro/banking-and-finance/sustainable-finance/overview-sustainable-finance_it; accessed on 28 February 2021.
8. According to J.P. Morgan's estimates the total value of assets following sustainable strategies in 2020 was estimated to be as high as €45 trillion (Morgan, 2020).
9. Proposal for a Regulation of the European Parliament and of the Council on establishing the framework for achieving climate neutrality and amending Regulation (EU) 2018/1999 (European Climate Law), European Commission COM/2020/80 final, 4 March 2020 and Amended proposal, COM(2020) 563 final, 17 September 2020.
10. Communication from the Commission to the European Parliament, the European Council, the Council, the European Economic and Social Committee, the Committee of the Regions and the European Investment Bank, a Clean Planet for all. A European strategic long-term vision for a prosperous, modern, competitive and climate neutral economy. Brussels, 28.11.2018, COM(2018) 773 final. Retrieved from https://eur-lex.europa.eu/legal-content/EN/TXT/PDF/?uri=CELEX:52018DC0773&from=EN.
11. https://eur-lex.europa.eu/legal-content/EN/TXT/?uri=CELEX:52018DC0773.
12. 'Communication: The European Green Deal', European Commission, COM/2019/640 final, 11 December 2019.
13. Proposal for a Regulation of the European Parliament and of the Council on establishing the framework for achieving climate neutrality and amending Regulation (EU) 2018/1999 (European Climate Law), European Commission COM/2020/80 final, 4 March 2020 and Amended proposal, COM(2020) 563 final, 17 September 2020.
14. 'Communication: Forging a climate-resilient Europe—the new EU Strategy on Adaptation to Climate Change', European Commission, COM/2021/82 final, 24 February 2021.
15. Proposal for a Directive of the European Parliament and of the Council amending Directive 2003/87/EC establishing a system for greenhouse gas emission allowance trading within the Union, Decision (EU) 2015/1814 concerning the establishment and operation of a market stability reserve for the Union greenhouse gas emission trading scheme and Regulation (EU) 2015/757, European Commission COM(2021) 551 final, 14 July 2021.

16. Proposal for a Regulation of the European Parliament and of the Council amending Regulation (EU) 2018/842 on binding annual greenhouse gas emission reductions by Member States from 2021 to 2030 contributing to climate action to meet commitments under the Paris Agreement, European Commission, COM(2021) 555 final, 14 July 2021.
17. Proposal for a Regulation of the European Parliament and of the Council amending Regulations (EU) 2018/841 as regards the scope, simplifying the compliance rules, setting out the targets of the Member States for 2030 and committing to the collective achievement of climate neutrality by 2035 in the land use, forestry and agriculture sector, and (EU) 2018/1999 as regards improvement in monitoring, reporting, tracking of progress and review, European Commission, COM(2021) 554 final, 14 July 2021.
18. 'Communication: New EU Forest Strategy for 2030', European Commission, COM(2021) 572 final, 16 July 2021.
19. Proposal for a Directive of the European Parliament and of the Council amending Directive (EU) 2018/2001 of the European Parliament and of the Council, Regulation (EU) 2018/1999 of the European Parliament and of the Council and Directive 98/70/EC of the European Parliament and of the Council as regards the promotion of energy from renewable sources, and repealing Council Directive (EU) 2015/652, European Commission, COM(2021) 557 final, 14 July 2021.
20. Proposal for a Directive of the European Parliament and of the Council on energy efficiency (recast), European Commission, COM(2021) 558 final, 14 July 2021.
21. Proposal for a Council Directive restructuring the Union framework for the taxation of energy products and electricity, European Commission, COM(2021) 563 final, 14 July 2021.
22. Proposal for a Regulation of the European Parliament and of the Council amending Regulation (EU) 2019/631 as regards strengthening the CO2 emission performance standards for new passenger cars and new light commercial vehicles in line with the Union's increased climate ambition, European Commission, COM(2021) 556 final, 14 July 2021.
23. Proposal for a Regulation of the European Parliament and of the Council on the deployment of alternative fuels infrastructure, and repealing Directive 2014/94/EU of the European Parliament and of the Council, European Commission, COM(2021) 559 final, 14 July 2021.
24. Proposal for a Regulation of the European Parliament and of the Council on ensuring a level playing field for sustainable air transport, European Commission, COM(2021) 561 final, 14 July 2021.
25. Proposal for a Regulation of the European Parliament and of the Council on the use of renewable and low-carbon fuels in maritime transport and

amending Directive 2009/16/EC, European Commission, COM(2021) 562 final, 14 July 2021.

26. Proposal for a Regulation of the European Parliament and of the Council establishing a carbon border adjustment mechanism, European Commission, COM(2021) 564 final, 14 July 2021.
27. 'Communication: Powering a climate-neutral economy: An EU Strategy for Energy System Integration', European Commission, COM/2020/299 final, 8 July 2020.
28. 'Communication: A hydrogen strategy for a climate-neutral Europe', European Commission, COM(2020) 301 final, 8 July 2020.
29. 'Communication: A Renovation Wave for Europe—greening our buildings, creating jobs, improving lives', European Commission, COM/2020/662 final, 14 October 2020.
30. Commission implementing regulation (EU) .../... detailing the technical modalities for the effective implementation of an optional common Union scheme for rating the smart readiness of buildings, European Commission C/2020/6929 final, 14 October 2020.
31. https://europa.eu/new-european-bauhaus/index_en.
32. 'Communication: EU Biodiversity Strategy for 2030 Bringing nature back into our lives', European Commission, COM/2020/380 final, 20 May 2020.
33. 'Communication: A Farm to Fork Strategy for a fair, healthy and environmentally-friendly food system', European Commission, COM/2020/381 final, 20 May 2020.
34. 'Communication: Sustainable Europe Investment Plan European Green Deal Investment Plan', European Commission, COM/2020/21 final, 14 January 2020.
35. Regulation (EU) 2021/241 of the European Parliament and of the Council of 12 February 2021 establishing the Recovery and Resilience Facility (*OJ L 57, 18.2.2021, pp. 17–75*).
36. 'Technical Guidance on the application of 'do no significant harm' under the Recovery and Resilience Facility Regulation' *(OJ C 58, 18.2.2021, pp. 1–30)*.
37. 'Communication on new funding strategy to finance NextGenerationEU', European Commission, COM/2021/250 final, 14 April 2021.
38. Proceeds from the issuance of green bonds are linked to projects aimed at achieving environmental sustainability objectives (e.g. pollution prevention and control). Recently, several additional categories of bonds proliferated, such as those named social bonds (linked to the achievement of social goals), sustainable bonds (referring to a combination of social and environmental objectives) and transition bonds, issued by carbon-intensive companies to finance projects for the transition to decarbonised business models.

39. In the EU, green bonds will be also used to finance 30% of the EUR 750 billion of the Next Generation EU plan (see State of the Union Address by President von der Leyen at the European Parliament Plenary of 2020, https://ec.europa.eu/commission/presscorner/detail/en/SPEECH_20_1655). In addition to green bonds, in the second half of 2020, three social bond issues were carried out within the framework of the European SURE (Temporary Support to mitigate Unemployment Risks in an Emergency) programme for a total of EUR 39.5 million: total issuance thus increased from EUR 23 million in the first half of the year to EUR 74 million at the end of the second half.

References

Beatley, T., & Manning, K. (1998). *The ecology of place: Planning for environment, economy and community*. Island Press.

Berke, R., & Conroy, M. M. (2000). Are we planning for sustainable development? An evaluation of 30 comprehensive plans. *Journal of American Planning Association, 66*(1), 21–33.

Campbell, S. (1996). Green cities, growing cities, just cities? Urban planning and contradictions of sustainable development. *Journal of American Planning Association, 62*(3), 296–312.

EBA. (2020a). *Sustainable finance—Market practice*. Staff Paper, 6. https://www.eba.europa.eu/sites/default/documents/files/document_library/Sustainable%20finance%20Market%20practices.pdf

EBA. (2020b). *Management and supervision of ESG risks for credit institutions and investment firms* (Discussion paper). https://www.eba.europa.eu/calendar/discussion-paper-management-and-supervision-esg-risks-credit-institutions-and-investment

EBA. (2021). *EBA report on management and supervision of ESG risks for credit institutions and investment firms*. https://www.eba.europa.eu/sites/default/documents/files/document_library/Publications/Reports/2021/1015656/EBA%20Report%20on%20ESG%20risks%20management%20and%20supervision.pdf

ESMA. (2020). *Trends risks vulnerabilities* 2. https://www.esma.europa.eu/sites/default/files/library/esma_50-165-1287_report_on_trends_risks_and_vulnerabilities_no.2_2020.pdf

ESMA. (2021). *Trends risks vulnerabilities* 2. https://www.esma.europa.eu/sites/default/files/library/esma50-165-1842_trv2-2021.pdf

European Commission. (2020). *Study on sustainability-related ratings, data and research*. https://op.europa.eu/en/publication-detail/-/publication/d7d85036-509c-11eb-b59f01aa75ed71a1/language-en/format-PDF/source-183474104

Eurosif. (2018). *European SRI Study 2018.* Retrieved from https://www.eurosif.org/wp-content/uploads/2018/11/European-SRI-2018-Study.pdf

Linciano, N., Cafiero, E., Ciavarella, A., Di Stefano, G., Levantini, E., Mollo, G., Nocella, S., Santamaria, R., & Taverna, M. (2021). *Sustainable finance: Trends, issues and perspectives amid the evolution of the EU regulatory framework* (Sustainable Finance Working Papers, No. 1). https://www.consob.it/web/consob-and-its-activities/abs_sf/-/asset_publisher/coLw917vXYH5/content/sustainable-finance-n-1/718268

Małgorzata, J., Pieloch-Babiarz, A., & Sajnóg, A. (2020). Does short-termism influence the market value of companies? Evidence from EU countries. *Journal of Risk Financial Management, 13*(11), 272. https://doi.org/10.3390/jrfm13110272

Morgan, J. P. (2020, May 18). *ESG and COVID-19: Friends or foes?* J.P. Morgan perspectives. https://insights.nordea.com/en/sustainability/covid-19-climate/

OECD. (2020). *OECD business and finance outlook 2020: Sustainable and resilient finance.* OECD Publishing, Paris. https://www.oecd.org/daf/oecd-business-and-finance-outlook-26172577.htm

OECD. (2021). *Transition finance: Investigating the state of play—A stocktake of emerging approaches and financial instruments* (Environment Working Paper No. 179). https://www.oecd.org/environment/workingpapers.htm

Pezzey, J. C. V., & Toman, M. A. (2005). Sustainability and its economic interpretations. In R. D. Simpson, M. A. Toman, & R. U. Ayres (Eds.), *Scarcity and growth: Natural resources and the environment in the new millennium* (pp. 121–141). RFF Press.

Porter, M. E., & Reinhardt, F. L. (2007). *Grist: A strategic approach to climate.* https://hbr.org/2007/10/climate-business-_-business-climate%5D

Schoenmaker, D., & Schramade, W. (2019). *Principles of sustainable finance.* Oxford University Press.

Soppe, A. (2004). Sustainable corporate finance. *Journal of Business Ethics, 53*(1–2), 213–224. https://doi.org/10.1023/B:BUSI.0000039410.18373.12

Sparkes, R. (2001). Ethical investment: Whose ethics, which investment? *Business Ethics: A European Review, 10*, 194–205.

State Street Global Advisors. (2017). ESG Institutional Investor Survey. *Performing for the future. ESG's place in investment portfolios. Today and tomorrow.* https://www.ssga.com/investment-topics/environmental-social-governance/2017/esg-institutional-investor-survey-us.PDF

Timpano, M., & Fedeli, F. (2019). La finanza per lo sviluppo sostenibile: un'analisi dello stato dell'arte. *Economia Italiana*, 2019(2).

UN Environment and World Bank Group. (2017). *Roadmap for a sustainable financial system.* A UN Environment – World Bank Group Initiative.

WCED. (1987). *Report of the World Commission on Environment and Development: Our Common Future*. United Nations Documents. https://www.are.admin.ch/are/en/home/media/publications/sustainable-development/brundtland-report.html

Zadeh, A. A., & Serafeim, G. (2018). Why and how investors use ESG information: Evidence from a global survey. *Financial Analysts Journal, 74*(3), 87–103. https://ssrn.com/abstract=2925310

CHAPTER 3

Information as a Driver of Sustainable Finance

Claudia Guagliano, Nadia Linciano, and Paola Soccorso

Introduction

In recent years the debate on the capability of financial markets to correctly incorporate the Environmental, Social and Governance (ESG) factors relevant to economic activities into asset prices has progressively gained momentum as one of the key drivers of sustainable development. In fact, efficient pricing of both ESG performance and exposure to ESG

C. Guagliano
European Securities and Markets Authority, Paris, France
e-mail: claudia.guagliano@esma.europa.eu

N. Linciano (✉)
Commissione Nazionale per le Società e la Borsa, Rome, Italy
e-mail: n.linciano@consob.it

P. Soccorso
Commissione Nazionale per le Società e la Borsa, Rome, Italy
e-mail: p.soccorso@consob.it

N. Linciano et al. (eds.), *Information as a Driver of Sustainable Finance*, Palgrave Studies in Impact Finance,
https://doi.org/10.1007/978-3-030-93768-3_3

risks of financial products and economic activities is essential to channel resources to sustainable projects while granting allocative efficiency.

Climate changes raise special concerns, as market failure to account for the potential impact of extreme events (so-called physical risk) and/or the potential effect of unexpected legislation accelerating decarbonisation (so-called transition risk) might prompt the abrupt repricing of 'exposed' financial assets. The 2015 Paris Agreement and the initiatives that followed are in fact accelerating the transition to a low carbon economy, in order to contrast the physical risks of climate changes.[1] Such acceleration makes an increasing number of carbon-intensive assets and processes become stranded. In the absence of 'smoothing mechanisms', this would prompt disruptive market reactions, resulting in a disorderly transition and affecting also the stability of financial systems (Board of Governors of the Federal Reserve System, 2020; Carney, 2018; ECB & ESRB, 2020).

At the same time, market prices should reflect not only the risks but also the opportunities arising from the integration of sustainability into financial actors' decision-making, for instance, accounting for the 'creative disruption' triggered by decarbonisation (OECD, 2021).

Finally, as attention to ESG investing grows, policymakers foresee the risk of an over-valuation of sustainability-related assets (ESMA, 2021), which could impair the efficient allocation of resources as severely as the underestimation of ESG features and exposure to climate changes would do.

To grant allocative efficiency and integrity of capital markets, a key precondition is the availability of high-quality sustainability-related data and information, enabling market participants to assess both the exposure to ESG risks and the ESG performance of businesses, projects and financial products.

However, raising the quality and the quantity of information may not be enough, as markets may still be unable to correctly price sustainability. In fact, the evidence gathered so far shows that prices incorporate the ESG-related information already publicly available with a delay, thus signalling informational inefficiencies. It is therefore key to understand the roots of these inefficiencies and whether they can be removed with adequate policy actions.

To delve deeper into this matter, the present chapter will first of all provide an overview of the ecosystem of sustainable information and of the contribution of its agents to the production and the dissemination of information in section "The Ecosystem of Sustainable Information". Section "Market Prices and ESG Factors" will review the empirical evidence on market pricing of ESG factors, with a focus on climate changes. Section "Why Markets May Fail in Pricing ESG Factors" will analyse and focus on the factors that may impair market pricing of sustainability-related risks and opportunities. Finally, section "From the Efficient Market Hypothesis (EMH) to the Adaptive Market Hypothesis (AMH)" will discuss whether sustainable investing and long-term value creation may be best framed within an asset pricing paradigm alternative to the efficient market hypothesis, compatible with fundamental analysis and active management.

The Ecosystem of Sustainable Information

The ecosystem of sustainability-related information basically encompasses the same categories of actors that play a role in the production and dissemination of the so-called financial information. It is key to the flow of resources from investors' willing to employ sustainable investment strategies to issuers and distributors of sustainability-related products and services and, ultimately, to the development of sustainable finance (Fig. 3.1).

Among market participants mainly generating and disseminating information, companies are the main producer as they gather data and information on the ESG characteristics of their business, the sustainability of specific activities and projects, and the ESG risks they are exposed to and the way they mitigate them. In the EU, pursuant to 2014/95/EU Directive, also called the Non-Financial Reporting Directive (NFRD), European large public-interest companies, with more than 500 employees, have to publish information related to environmental matters, social matters and treatment of employees, respect for human rights, anti-corruption and bribery, diversity on company boards (in terms of age, gender, educational and professional background; for more details see Chapter 4). This covers approximately 11,700 large companies and groups across the EU, including listed companies, banks, insurance companies and other companies designated by national authorities as public-interest entities. Firms may also deliver information upon request directly to other institutions, such as data providers, ESG rating/scoring agencies,[2] investors and other sustainability-related product and service providers offering research, benchmarks, voting advisory, etc. In fact,

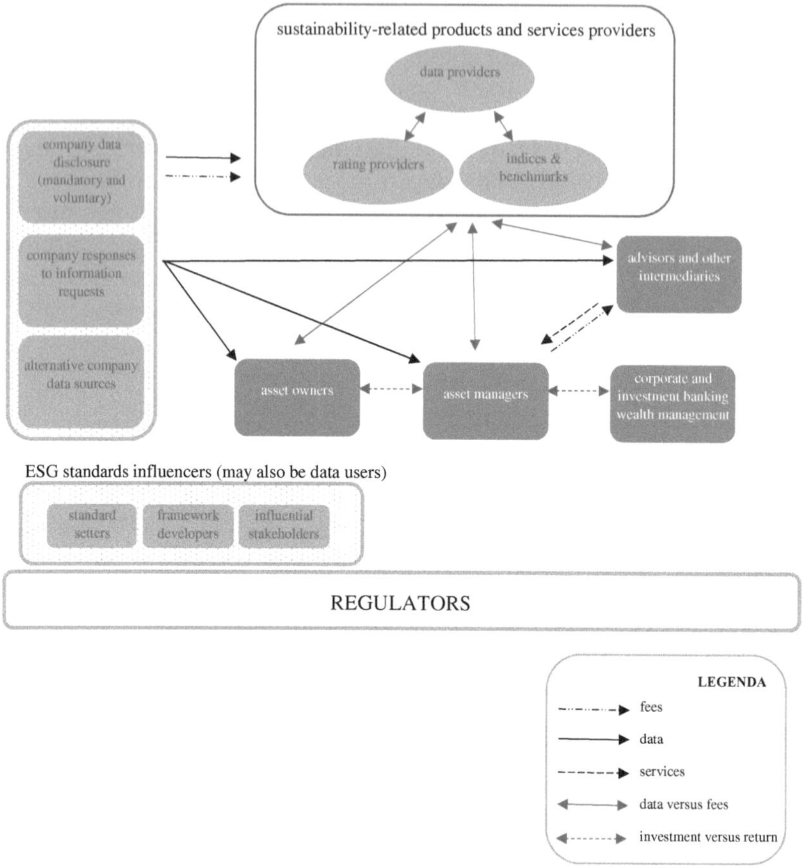

Fig. 3.1 The ecosystem of sustainable information (*Source* Authors's elaboration on European Commission [2020])

sustainability-related information is disseminated also by these institutions.

According to some studies cited in European Commission (2020), worldwide the number of ESG rating and data providers approximately amounts to 150, while those domiciled in the EU range between 30 and

40.[3] Ratings may focus either on firms to assess their sustainability performance or on financial products, such as a specific bond or a fund, to assess its ESG performance (for more details, see Chapter 5).[4]

Among market participants mainly using information, the broad category of investors encompassing asset owners and asset managers employ ESG information to make portfolio decisions and engage with company boards, also depending on investment strategies (for more details, see Chapter 6). Another key actor is represented by investment banks that assess sustainability-related data and information to make buy, hold and sell recommendations to investors. Based on the statistics of the signatories of the Principle for Responsible Investment (PRI),[5] as of July 2020, there are 552 asset owner signatories to the PRI, of whom 234 are EU 27-based, and 2348 asset manager, 760 EU 27-based, while EU-based investment banks are estimated to be 142. Signatories to the PRI include also professional advisors and consultants, i.e. professional service providers and advisers to all other market participants, for example, in relation to the preparation and presentation of company sustainability-related disclosures. As of September 2020, there are 350 service providers of whom 103 EU 27-based (Foubert, 2020) (Fig. 3.2).

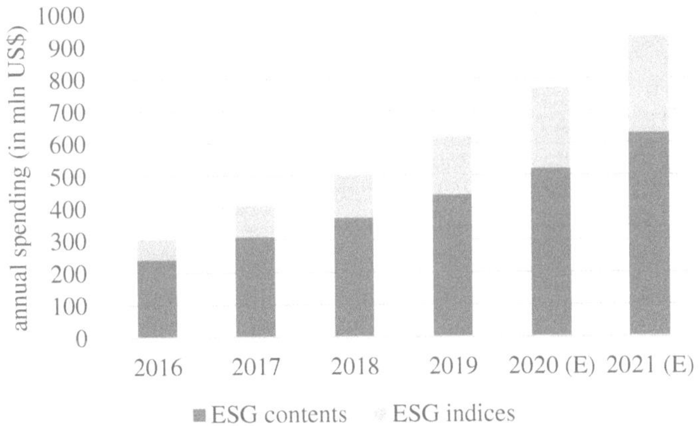

Fig. 3.2 ESG data market size (*Source* Foubert [2020] with Opimas estimates. Retrieved from http://www.opimas.com/research/547/detail/)

Additional providers and/or users of ESG data include benchmark administrators, offering indexes to measure the performance of investment funds,[6] and sell-side research and financial analysts that are estimated to be 463 firms at the global level, of which 55% domiciled in continental Europe.

In the ecosystem of sustainability-related information, bodies developing both standards and frameworks can be regarded as facilitators, as they contribute to enhance reliability and comparability of the information disclosed by companies and, ultimately, to the better functioning of the ecosystem itself. Among standard setters, well-known examples include the Climate Disclosure Standards Board, the Global Reporting Initiative (GRI) and the Sustainability Accounting Standards Board (SASB) as well as industry associations and other bodies also publishing guidelines on sustainability-related reporting for companies (European Commission, 2020). Among framework developers, whose focus is mainly on principle-based concepts of reporting, examples include the International Integrated Reporting Council (IIRC), the UN Global Compact, the GRI 10 Reporting Principles and the TCFD (more on this in Chapter 4).

Finally, regulators and financial supervisors intervene in the sustainable investment market through the regulation of market participants and their activities as well as the supervision and enforcement of the regulation. As for the European framework, this will be expanded on further in Chapter 10.[7]

Overall, total spending on 'ESG data' globally reached USD 617 million in 2019, and it is estimated to amount USD 745 million at the end of 2020 (Foubert, 2020). Europe accounts for 60% of this spending, and this share is likely to rise further following the implementation of the ongoing regulatory developments. As for the distribution of spending by segments of market participants, asset managers account for around 59% of the global market for 'ESG data', followed by sell-side institutions (19% of total market spend), asset owners (around 12%) and consulting firms, financial and other investment advisors (around 6%) (Foubert, 2020).

Market Prices and ESG Factors

Market Pricing of ESG Factors

The development of sustainable finance rests on financial markets' capability to factor into market valuations of economic activities and financial assets their ESG characteristics also in terms of exposure to ESG risks, to the benefit of market participants willing to integrate sustainability in their decision-making process.

First, it is useful to focus on the relation between ESG pricing and corporate social responsibility (CSR).

The empirical evidence available on this point, coming both from market studies and academic work, is ambiguous, due also to lack of data and methodological shortcomings.

Clark and Viehs (2014) and Clark et al. (2015) highlight the fact that studies typically focus on the impact of single factors (either E or S or G) without considering the multi-dimensional nature of CSR.

Schoenmaker and Schramade (2019) argue that many studies failed in disentangling the impact of ESG performance on each value driver that in turn affects stock performance. On this point, there are a number of studies identifying the channels through which the integration of ESG factors may create value (Clark et al., 2015; Kurucz et al., 2008; PRI-UN Global Compact, 2013): costs reduction (also in terms of cost of equity and cost of debt; Cremers & Nair, 2005; Cremers et al., 2007) and risk mitigation (also reputational and legal risks; Bauer & Hann, 2010); maximisation of profit and performance (in terms of revenues from process and product innovation; Porter & Kramer, 2006) and increasing reputation (among workforce, consumers, etc.; Roberts & Dowling, 2002).

However, the studies exploring the relationship between ESG and corporate financial performance (CFP) may suffer from potential endogeneity. Good ESG performance goes hand in hand with satisfactory economic results, which can be explained in two ways: '*either more responsible firms tend to be more profitable or, alternatively, more profitable firms tend to channel more resources into projects that increase the wellbeing of stakeholders*' (Kruger, 2015).

In addition, equivocal evidence on the relationship between CSR and CFP may be due to measurement issues pertaining to both CSR and CFP. Galant and Cadez (2017) review alternative measurement approaches for both concepts, spanning from reputation indices, content analyses, questionnaire-based surveys and one-dimensional measures as for the CSR and using accounting-based measures, market-based measures and combined measures as for CFP.[8] Some authors investigate the performance of the ESG ratings and scores as a measure of the ESG performance of businesses, although they have proved to suffer from many drawbacks in terms of methodological opaqueness and inconsistencies (see Chapter 5). In particular, Khan et al. (2016) analyse the relation between the materiality of ESG issues (as assessed by applying SASB principles) and the information content of ESG ratings,[9] and find that only firms with good ratings on material sustainability issues significantly outperform firms with poor ratings on these issues, while firms with good ratings on immaterial sustainability issues do not significantly outperform firms with poor ratings on the same issues. These results highlight that accounting standards may help in identifying ESG issues that are material and as such affect financial performance (as for the notion of materiality, see Chapters 4 and 9). Quoting Eccles and Serafeim (2013): '*It's possible to simultaneously boost both financial and ESG performance if you focus strategically on issues that are the most "material" to shareholder value, and you develop major innovations in products, processes, and business models that prioritize those concerns*'.

Among the studies investigating the relationship between CSR and CFP measured in terms of accounting-based indicators, meta-analysis of a significant number of econometric academic studies supports the hypothesis of the complementarity between profitability and social responsibility (Clark et al., 2015; Friede et al., 2015). Findings from the meta-study by Clark et al. (2015) suggest that firms with strong sustainability scores show better operational performance and are less risky. Friede et al. (2015) reviewed more than 2000 empirical researches, mostly reporting evidence about a positive and stable-over-time ESG impact on corporate financial performance. Empirical results by Wu and Shen (2013), referring to 162 banks in 22 countries, show that CSR is positively related to financial performance, as it is measured by accounting-bases measures (in terms of return on assets, return on equity, net interest income and non-interest income). The experimental study by Greening and Turban (2000) points out that superior CSR practices might attract high-quality workforce, which then potentially prompts higher productivity and efficiency and better operational performance. After analysing data from 121 US-based matched-pairs (242 individual firms) over a 15-year period,

Ortiz-de-Mandojana and Bansal (2016) conclude that social and environment practices contribute to companies' organisational resiliency lowering financial volatility, facilitating access to resources and improving their degree of adaptability and survival rate. Eccles et al. (2014), using a matched sample of 180 US companies, find that high sustainability firms are more likely to be more long-term oriented and to significantly outperform their counterparts over the long-term, both in terms of accounting as well as stock market performance.

This latter result call for the consideration of another strand of literature investigating the relationship between corporate governance attaching importance to ESG factors, on one hand, and market-based indicators (stock prices and corporate bond spreads), on the other.

Clark et al. (2015) confirm that sustainability is valued by the stock market, i.e. more sustainable firms generally outperform less sustainable firms. Nagy et al. (2016) find that the performance of a portfolio incorporating ESG strategies outperformed that of the MSCI World index between 2007 and 2015 and, since they analyse stock returns of two different strategies, conclude that a significant part of this outperformance was not explained by management style, and thus may have been attributable to ESG factors. Schoenmaker and Schramade (2019) point out that, since ESG performance is often proxied by ESG ratings which are quite static, the impact on stock returns may be better signalled by changes in ratings rather than by rating levels. Results by NN Investment Partners and ECCE (2017) find that companies with improving ESG ratings outperform the stock market, while already high-rated companies do not. Also findings by Khan et al. (2016) are based on changes in ESG performance rather than on levels of ESG performance.

As for the impact on equity and debt cost, the review study by Clark and Viehs (2014) supports the hypothesis that the risk-reduction characteristics of proper CSR standards lowers the cost of equity financing because of a reduced firm-specific risk, while Clark et al. (2015) conclude that good corporate governance structures, good environmental management practices and employee well-being reduce borrowing cost. Polbennikov et al. (2016) find that higher-ESG-rated corporate bonds have slightly lower spreads, all else being equal, and modestly outperform their lower-rated peers when controlling for various risk exposures. Bauer and Hann (2010) investigate more than 2200 bond issues in the US regarding the effects of corporate social responsibility quality and its effects on bond spreads documenting a significant and negative relationship between good environmental management practices and a firm's loan spread. Capelle-Blancard et al. (2017) analyse the extent to which ESG performance can affect borrowing cost in international capital markets, finding that well-ESG-performing countries tend to have a lower risk of default and thus a lower bond spreads. Moreover, the economic impact is stronger in the long run and, in a cross-country comparison, more significant in the euro area than in other OECD countries.

Market Pricing of Risks and Opportunities of Climate Changes

Climate changes driven by greenhouse gas emissions are a top issue in the policy agenda. A growing strand of the literature and of the policy debate is focusing on whether financial markets can price the financial risk descending from climate changes. The issue is challenging as it refers to two categories of risk that could move in different directions: physical risk and transition risk.[10]

Physical risk is driven by climate alterations impacting human, natural and physical capitals and impairing the economic and financial resilience of market participants (Box 3.1).

Box 3.1: Physical Risk

Physical risk is associated to the effects driven by climate alterations impacting human, natural, and physical capitals and impairing the economic and financial resilience of market agents (firms, financial intermediaries, households, and governments). Physical risk can be classified into acute and chronic risks. Acute risk refers to extreme events, such as heatwaves and floods growing in frequency and severity, while chronic risks refer to structural and long-term effects of climate changes, such as drought and rising sea levels. In particular, physical risk can be identified through the following seven features: increasing, spatial (i.e. varies across geographical areas), non-stationary, non-linear beyond given thresholds, systemic, and regressive as it raises inequalities across countries, and unexpected as firms and communities efforts to manage climate changes are likely to result insufficient (McKinsey Global Institute, 2020).

The devastating impact of the twofold dimension of realised physical risks, acute and chronic, is already blatant in many areas across the world and is forecast to intensify further as time goes by, especially in the absence of decarbonisation.[11]

Available estimates point to a physical damage from climate hazard up to one-tenth, or even one-fifth, of global GDP by the end of this century, which in terms of current global output would amount to USD 8–17 trillion. In Europe, under a high warming scenario (e.g. under a temperature rise above 3 °C compared to the pre-industrial period) and in the absence of planned or public adaptation, the overall welfare loss is estimated to be around 1.9% of GDP (240 billion euros) per year by the end of the century (Ciscar et al., 2018). This figure is likely to be underestimated, as it regards only six of the 11 relevant impact areas of climate change identified in the study (coastal floods, river floods, droughts, agriculture, energy, transport, water resources, habitat loss, forest fires, labour productivity, and mortality due to heat). In addition, the Mediterranean area is the most exposed in many areas, in a clear geographical north–south divide. Most of the welfare losses would be greatly reduced under a 2 °C scenario.

Physical risk in climate-related events can affect macroeconomic conditions and financial stability through different channels (NGFS, 2019). First, they erode the collateral and asset values of insurers (Albouy et al., 2013; Bunten & Kahn, 2014; Gassebner et al., 2010). In the EU insurance sector, weather-related costs already account for over 80% of catastrophe-related losses. At the same time, an increasingly insurance protection gap is mounting: weather-related losses in the EU amounted to €537 billion between 1980 and 2018 and only 35% of these were insured (ECB & ESRB, 2020). Climate risks may also exacerbate existing financial system vulnerabilities, such as historically high levels of corporate leverage. In addition, disruption in business activity and trade, and destruction of infrastructure can divert capital from technology and innovation to reconstruction and replacement (Batten, 2018). Social conflicts may be exacerbated by migration flows triggered by climate changes (Opitz Stapleton et al., 2017). On the demand side, household wealth and therefore private consumption could be hit by climate changes, and this effect can be reinforced further by the reduction in the collateral values of housing stock, while a fall in consumption can contribute to dampen business investments (Hallegatte, 2009).

A fully-fledged estimate of the socioeconomic consequences of climate changes should take into account that the amount of physical risk that manifests also depends on the adaptation measures that might be put in place to contrast the climate hazards and that in turn reduce risk.[12] Still, even after considering potentially mitigating responses to climate changes, the effects on economic and financial systems remain significant.

The decline in asset values due to materialised physical risks impairs not only the business continuity of the firms hit by a catastrophic event either directly or indirectly (e.g. through the supply chain) but also the stability of exposed financial institutions. The erosion of collateral and asset values can harm a large number of insurance companies, as their liabilities rise, and banks, as their credit quality worsen. Giuzio et al. (2019) highlight that the share of weather-related catastrophe losses has steadily risen over time hitting over 80% of insured catastrophe losses in 2018, along with a growing frequency of weather-related loss events recording a new record at the end of the same year.

Transition risk (also adaptation risk) descends from unanticipated changes in the pace of decarbonisation of economic activities and of fossil-fuel usage, triggered by technological innovation, changes in investors' preferences and tighter climate legislation (Box 3.2).

Box 3.2: Transition Risk

Transition risk refers to unanticipated changes in the pace of decarbonisation of economic activities and fossil-fuel usage. These changes may be triggered by several factors: disruptive technological innovation, shift in investors' preferences, legal actions, and tighter climate legislation. Therefore, climate-related transition risk includes the technology risk concerning the uncertainty in technology developments and deployment (e.g. lowering the cost of production of renewable energy), market risk linked to adverse dynamics in the demand of given products (e.g. higher demand for renewable energy), reputational and litigation risks (e.g. arising in the context of climate-related damages), credit risk (e.g. higher default risk of carbon-intensive firms—resulting eventually into deterioration of the quality of bank exposures—and higher sovereign risks for countries with carbon-intensive industries), and policy risks linked to a heightened commitment of policymakers to environmental goals (e.g. new limits to greenhouse gas emissions or carbon taxes). An additional source of transition risk may be the mispricing of fossil fuels and of the externalities of the associated carbon emissions. In fact, such an under-pricing feeds not only physical risk (as related to climate and weather events) but also transition risk, as abrupt adjustments in fossil fuel prices might trigger sharp corrections of market evaluation of the most carbon-intensive activities. According to Bank of England and NGFS estimates, credit losses from an unanticipated repricing of carbon could range between USD 1 trillion and 4 trillion on fossil fuels (OECD, 2021).

Carbon pricing may result either from a tax levied on emitters on each metric ton of CO_2 emitted or from trading on a market of permits to pollute (such as the European Union Emissions Trading System, launched in 2005 as one of the pillars of the EU energy policy) or from an increasing corporate practice known as internal carbon pricing.[13] Current carbon prices are however too low and too heterogeneous across firms and areas[14] relative to the \$40–80 per metric ton of CO_2 emissions recommended for 2020 to meet the 'well below 2 °C' temperature goal of the Paris Agreement (World Bank, 2021). As carbon prices in the recommended range cover less than 5% of global emissions, alternative policy measures are advocated to accelerate the green transition (e.g. removal of fossil fuel subsidies, ban on carbon-intensive technologies and stimuli to innovations, new business models, and markets; Rosenbloom et al., 2020).

If transition risk materialises, assets (such as resource reserves, infrastructure or industries) may quickly become obsolete thus suffering from premature write-offs, downward revaluations or conversion to liabilities, turning into increased costs, losses and climate-related asset stranding.[15] So far, this issue has received limited attention by researchers, in spite of the potentially disruptive impact on energy resources[16] and the potentially severe consequences that may materialise in the near future. Especially developing countries, disproportionately dependent on fossil fuel extraction and exports, may suffer in terms of lower growth,[17] while developed economies heavily reliant on coal-related revenues may experience an erosion of their tax base and budgets triggering adverse impacts on public spending.[18] In addition, the distributive effects of the green transition may foster a trade-off with other SDGs, such as the eradication of poverty (Campagnolo & Davide, 2019). On the other hand, successful transitions to greener economies may eventually benefit countries by feeding a stable and resilient growth, providing opportunities for new industries, strengthening the institutional environment and alleviating poverty (Ansari & Holz, 2020). OECD (2019) estimates that meeting the goal of the Paris Agreement could translate into an increase in global GDP by up to 5 percentage points.

An essential precondition of informed policy decisions is the quantification of both physical and transition risks. However, any measurement and estimate exercise suffer from uncertainty about climate changes over the long-term, as well as from lack of data on the impact of climate risks on individual companies and financial markets (banks, insurance companies and securities markets),[19] and limitation in modelling that should integrate in a cross-disciplinary fashion traditional macro-financial approaches with climate science analysis. The interplay of physical and transition risks is complex and requires a simultaneous advance both in climate mitigation and climate adaptation measures (CFTC, 2020). In fact, a trade-off may exist but is due to vanish in the long run as adaptive policies heighten transition risks in the short term while reducing physical risk in the long term NGFS (2020b).[20] In some areas, such a trade-off may not show up: in the real estate, for instance, assets valuation may suffer at the same time from both absolute losses due to physical risks and the transition risk of consumer preferences dismissing assets that lack sustainability. This example highlights that exposure and resilience to physical and transition risk do not conform to clear patterns and that the assessment of climate risk requires an in-depth analysis both at sectoral and company and asset levels (S&P Global, 2019). In addition, timing of intervention is key, as late actions could in fact be associated with a direct, self-reinforcing relationship between physical and transition risks if abrupt tightening is put in place to compensate physical disruptions due to previous delays. This could also lead to a sharp reassessment of climate change risks, which in turn would trigger a financial market reassessment feeding a persistent tightening of financial conditions as losses supervene (Bolton & Kacperczyk, 2021).

In the Euro area, first estimates of the financial institutions' exposures to transition risk have mainly regarded specific industrial sectors, typically selected in the NACE classification[21] on the basis of aggregate environmental metrics (e.g. a measure of carbon emissions for the sector).[22]

Preliminary stress test exercises show that acting early is clearly beneficial to economic and financial systems (see, among the others and for further references, ECB, 2021; ECB & ESRB, 2020; ESRB, 2016; Bowen & Dietz, 2016). A preliminary economy-wide climate stress test run for the Eurozone shows that, absent further climate policies, extreme weather events would be a significant source of systemic risk (ECB, 2021).[23] Heatwaves and wildfires, for instance, would raise the probability of default of companies most exposed, which in turn would impair the financial stability of those banks with portfolios concentrated in vulnerable economic sectors and geographical areas. In this framework, the short-term costs of a governed transition are disproportionally lower than the medium-long term costs of unmanaged climate changes. In addition, actions directed at transition trigger investments in more efficient technologies thus contributing to enhance firms' performance and market valuations. This latter effect, somehow neglected in the analyses of the impact of transition, strengthen the need of a rapid shift towards a greener economy, not only to meet the targets of the Paris Agreement but also to manage and mitigate the disruptive effects of climate changes.

Do financial markets efficiently price climate risks? Evidence gathered so far with respect to both physical and transition risks is mixed, due also to some methodological challenges.[24] However, the evidence generally points to a reaction of market prices that is increasing over time, probably as the result of a learning process, although reflecting behavioural distortions such as salience bias and limited attention.

As for physical risk, Bansal et al. (2016) use data from 39 countries and from the US capital markets from 1970 and 2012, when almost all economies experienced a significant increase in temperature, and show that a one degree Celsius rise in temperature is associated with a decline of about five percentage points in equity valuations, and that such negative impact has been rising over time. Hong et al. (2018) use data from thirty-one countries with publicly traded equities in the food industry and find that stock prices respond with a significant delay to trends in droughts that will likely drive relatively poor subsequent profit growth. Kumar et al. (2019) find that for US-listed firms exposure to temperature changes predicts stock returns, thus suggesting that stock markets underreact to corporate climate sensitivity. Shashwat et al. (2020) find evidence of mis-estimation of climatic disasters by US fund managers due to a salience bias, that is the tendency to overweight probabilities of climate hazards depending on the ease with which events can be recalled. Such bias leads fund managers to underweight stocks recently exposed to extreme climatic events in their portfolios and fades away over time and distance from the disaster. Xu et al. (2021) confirm findings by previous studies (in particular, Hong et al., 2018) by showing that, over 1970–2019, stock prices of firms listed on the New York Stock Exchange (NYSE), Nasdaq, and American Stock Exchange (AMEX) vary with the climate change exposure of firms (as resulting from public information available to all investors) in a predictable manner. The authors explain market inefficiency with investors' little attention to salient information on firm's exposure to climate risk, which also drives a return predictability more severe in more acute states of climate change.

As for financial markets' reaction to transition risk, preliminary evidence shows that prices may factor in such risk although not to a full extent yet. Atanasova and Schwartz (2019) analyse the relation between the value of a sample of 600 North American oil companies over 1999–2018 and their developed and undeveloped reserves and document that future investment in undeveloped reserves growth is associated with a negative impact on firm value, which is stronger in countries with strict climate policies where reserves may become stranded assets in the near future. Choi et al. (2020) use data on 74 exchange cities from 2001 to 2017 and find evidence that carbon intensive firms earn lower stock returns when the local exchange city is abnormally warmer. These patterns, which seem persistent in the longer term, are consistent with experiential learning as people update their beliefs on global warming when experiencing warmer than usual temperatures and review their investment choices accordingly. Additional evidence on investors' efforts to explore how to manage climate risks in their investment portfolios is brought by Engle et al. (2020) and Bolton and Kacperczyk (2021), using returns on stocks of US-listed companies from 2005 to 2017 show that fund managers screen potential investments for direct carbon emissions and demand compensation for the associated transition risks. Bua et al. (2021) develop a methodology to disentangle market reactions to transition and physical risk by using media news and find that over 2015–2019 the equity returns of the Eurostoxx 600 constituents seem to have somewhat incorporated transition risk, implying that investors have sold brown assets most exposed to events accelerating the transition to a net-zero carbon economy. Amzallag A. (2021) finds that the EU funds whose portfolios are tilted towards more polluting assets (brown funds) distribute their portfolio over a larger number of companies than funds with cleaner portfolios (green funds). This apparent diversification hides a concentration risk: brown funds are more closely connected with each other (have more similar portfolios) than green fund portfolios, which tend to 'herd' less (have less similar portfolios to those of other green funds). This suggests that widespread climate-related financial shocks are likely to disproportionately affect brown funds.

Recent data show that since 2019 the median Price-Earnings (P/E) ratio of 'green' firms exceeded that of 'brown' firms. Higher P/E ratios indicate that investors are betting on future growth of these companies, supported by an apparent shift among many governments across the planet. Still ROE differential between green and brown companies remains very limited making 'green' firms 'particularly vulnerable to changes in sentiment' (ESMA, 2020, 2021). There is evidence that firms with high levels of carbon emissions have higher equity price returns, consistent with investors demanding a premium for bearing climate-related risks.[25]

Other studies find that firms engaged in the extraction of fossil fuels have recently faced a higher cost of funding.[26] With respect to sovereign exposures to climate-related risks, there is some evidence that investors demand a premium to hold sovereign bonds more exposed to physical risks (Cevik & Jalles, 2020).

There is, however, little evidence that broader market prices incorporate risk premia commensurate with the scale and nature of climate-related risks across different sectors. For example, IMF (2020) finds that, in aggregate, equity prices do not reflect the potential impact of the future crystallisation of physical risks under various adverse climate change scenarios.

Why Markets May Fail in Pricing ESG Factors

A very important question from policy perspective is whether there are sufficient market-based incentives that can make investors place their capital with less-polluting companies. There is some preliminary evidence showing that a growing number of investors, including large institutional investors and asset managers, may well be willing to receive lower returns in exchange for investing in more sustainable companies (Pastor et al., 2019). However, the assessment of ESG factors in investment decision will be stronger if firms that pollute less are consistently more profitable for investors than polluting companies.

Given the evidence reviewed so far, there are a number of factors which may make the assessment of ESG factors more complicated, including the following:

1. Lack of data/information: corporate disclosure and available information remain insufficient, incomplete and inconsistent. To enable market participants to incorporate ESG risks more effectively in their decisions, it is key to address the existing data gaps (FSB, 2021). In particular, the monitoring and the assessment of ESG risks require data able to capture exposures of financial firms to ESG-related risks and to perform a global comparison and aggregation of financial firms' exposures to ESG-related risks.

2. Short-termism. Current market practices often prompt market participants to focus on short-term performance rather than mid- to long-term objectives. It has been highlighted that short-termism constitutes a 'vicious cycle' in which the setting of short-term goals and metrics by companies in response to investor demand further contributes to shortening investors' horizons (Barton, 2017). A frequently cited reason for this vicious cycle is the fact that stock markets make forecasts of firm value based on companies' reported earnings, thus introducing a short-term or myopic incentive in company behaviour (Stein, 1989). Similarly, some have referred to a 'perfect storm' (2 De grees Investing Initiative, 2017) to describe a situation in which factors such as relatively flexible regulatory requirements, a race to the bottom among peer companies, a limited role of auditors, the fear of litigation and limited demand from financial analysts discourage long-term risk disclosures, thus fostering short-termism. In order to lessen some of the pressure for short-termism in financial markets, it has been suggested that the quality of the investor-corporate dialogue needs to be more long-term oriented (UN Global Compact, 2017). As of August 2021, over 2700 organisations with almost USD 100 trillion in assets under management were signatories to the Principles for Responsible Investment (PRI), indicating a commitment to incorporate sustainability issues into investment analysis and decision-making processes. ESG factors increasingly influence the allocation and monitoring of assets at major institutions, as highlighted by Deutsches Aktieninstitut and Rothschild and Co Deutschland (2018). Short-termism may be reinforced by some practices: Quarterly financial reporting by companies, variable pay systems based on annual results; monthly or quarterly benchmarks for measuring investor performance; marking-to-market investments; supervisory treatment of illiquid assets (Schoenmaker & Schramade, 2020, pp. 90–94; Stein, 1989). The misalignment of investment horizons between investors, asset managers and asset owners is one of the main indicators of short-termism in financial markets. While asset managers typically have a short-term horizon (1 year or less) in their asset evaluations and incentives, investors and asset owners may have much longer-term horizons. A short-termism 'vicious circle' has also been highlighted whereby the setting of short-term goals and metrics by companies in response to investor demand contributes

to shorten investors' horizons further. Based on a survey done by ESMA (2020), sell-side investment research and the remuneration of fund managers and executives are identified as potential factors determining excessive focus on short-term results. Improvements to the quality of ESG disclosure and institutional investor engagement can help investors take more long-term investment decisions.

3. Complexity and lack of sufficiently encompassing and rigorous scenario analysis. A full assessment of the degree and combination of ESG-related risks embodied in asset prices also entails methodological challenges. This assessment requires a range of ESG-related risks to be embedded in asset valuation models. However, such models face complex methodological challenges, e.g. the need to model the actions and expectations of economic agents over a long time period, as well as the interactions between physical and transition risk and technological change. Scenario analysis is still under development. To be effective in making comparable and consistent assessments of climate-related risks across the global financial system, having common scenarios for use by financial authorities in different jurisdictions is needed. Until recently, there was no consensus on meaningful global scenarios for use in such analysis. The set of global reference scenarios established by the NGFS provides a common starting point for central bank climate risks scenario analysis (NGFS, 2020a).
4. Markets do not even use the already available public information/limited attention: Investors pay insufficient attention to ESG risks, or struggle to incorporate them in their pricing of risk due to their complexity. Financial market participants have traditionally used data and developed methodologies to assess financial risk. Even if the amount of available information and data is increasing, there is still no taxonomy agreed at global level to be used for ESG activities. In addition, further complexities derive from the need to take into account very long time horizons, in a context of elevated uncertainty, also related to technological and policy developments (Giuzio et al., 2019), without mentioning data gaps that, as already highlighted, are still relevant and impair a full assessment. Finally, the increasing role of high-frequency trading[27] in secondary markets and its impact on the price discovery process[28] may also reduce the capability of financial actors and investors to correctly price ESG risks.[29]

5. Limited comparability of ESG ratings. ESG ratings are scores assessing entities on criteria related to ESG themes of relevance for business or society (or both). The concept covers a wide variety of ratings, as at present there is no legal definition of what constitutes an ESG rating (Mazzacurati, 2021). When looking at the market one can broadly distinguish three main categories. Most ESG ratings try to capture the exposure of entities to ESG risks and the extent to which these risks are managed. The second category of ratings measures the impact that entities have on ESG factors, such as air pollution or gender (im)balance in key positions. The last type of ratings looks at very specific aspects, such as the degree of ESG disclosure, or their relevance for credit ratings. Not all ratings cover 'E', 'S' and 'G' factors; instead some focus on only one of the three. The object of ESG ratings is usually a private-sector or public-sector issuer rather than a financial instrument. ESG ratings have become increasingly popular and receive a fair amount of attention from the media and academics, who often criticise the limited comparability of such ratings. For example, a study finds that ESG ratings across five major ESG rating providers were only 60% correlated, compared with 99% for credit ratings from the three largest CRAs. This mainly reflects differences in the measurement of variables and aggregation methodologies. Adding to this problem is the lack of transparency of methodologies. To palliate ESG data gaps and consistency issues, some sustainability rating agencies rely on third-party providers or proprietary sources, meaning that some of the data and assumptions underlying ESG ratings are not freely accessible by investors. A third problem is the difference between relative ratings (positioning entities among their sector peers) and absolute ratings (comparing all rated entities on the same scale). Lack of clarity about this distinction can leave investors wondering why, for example, oil companies can receive higher ratings than firms that are active in cleaner industries. Such differences in the definition, scope and methodology used by ESG rating providers cause confusion among investors.

From the Efficient Market Hypothesis (EMH) to the Adaptive Market Hypothesis (AMH)

The EMH claims that asset prices incorporate all the information relevant to investors. Depending on the set of information referred to, the EMH can hold in the weak form (only past information is accounted for into current prices), semi-strong form (current prices reflect all publicly available information) and strong form (all information, regardless of whether it is public or not, is impounded).[30]

The EMH, at the basis of the modern portfolio theory (MPT), leads to the conclusion that passive investing is the best strategy. In this framework, the market portfolio refers to the allocation defined on the basis of prices reflecting fundamental values, i.e. incorporating all relevant information and there is no added value to investors from sketching future scenarios for economies, sectors or firms.

Over the last twenty years, the asset management industry has experienced a substantial shift from active to passive strategies, such as indexing, i.e. strategies that track an index by holding all of its constituent assets. This phenomenon has risen concerns on many grounds, such as industry concentration, asset prices, volatility, price discovery, and market liquidity (Kenechukwu et al., 2020). Some authors claim that index investing raise co-movements in asset prices, blunt stock price signals and impair price discovery, given that buy and sell decisions are not based on fundamentals but rather on benchmark index weights, reflecting the real-time prices prevailing on the market (DeLisle et al., 2015). In other words, passive investing by its nature could exacerbate short-termism (with consequences extensively discussed in section "Why Markets May Fail in Pricing ESG Factors"), as it focusses only on volatility risks, while neglecting fundamental corporate risks including those linked to ESG factors. Therefore, according to the critics, the EMH from '*an idealised way of describing market behaviour based on rational expectations economics has ironically become the instrument by which irrational behaviour becomes institutionalised*' (ILG, 2014). In fact, investors' irrational behaviour and the use of heuristics (Fama, 1998; Kahneman, 2013 among others) are the main explanation of the many anomalies of the EMH, detected by an extant strand of the literature (some authors claim the EMH is '*empirically false or at best a truism (i.e. true by definition, not by observation)*'; ILG, 2014). When considering, in particular, the pricing of environmental and social risk and the integration of all the information concerning ESG issues, as shown in the previous paragraph, many factors beyond irrationality and short-termism may play a role, such as the absence or the shortage of reliable, homogeneous and comparable sustainability-related data and information. That being said, the question is whether it is necessary to frame sustainable investing within a theoretical framework alternative to the EMH, in order to foster a long-term value-creation approach. Schoenmaker and Schramade (2020) advocate the application of the Adaptive Market Hypothesis (AMH), accommodating irrational behaviour and claiming that in line with an evolutionary perspective the information efficiency of markets evolves over time as a function of the number and quality of market participants who are sensitive to new information (Lo, 2017). Consistently with the AMH framework, the incorporation of ESG information into stock prices is an 'adaptive process', which would not be completed since not enough investors learn about the impact of ESG risks on companies' risks and profits and take them into account in their investment choices. This in turn is made possible by an active investment approach, based on fundamental analysis assessing companies' social and environmental value, alongside their financial value, distancing from short-term metrics and benchmarks, leveraging on shareholders' engagement to stir the integration of ESG factors to corporate decision-making processes. As non-financial information becomes more reliable and standardised and fundamental analysis more accurate and widespread, conditions will become increasingly favourable to active management and sustainability information will be increasingly incorporated into market prices along with all available information. Market valuations will signal the value of all factors that contribute to the value of the company as a whole, in a 'new vision of value' (KPMG, 2014) and in a new vision of information.

Notes

1. As recalled in Chapter 2, the Paris Agreement is a legally binding international treaty on climate change, adopted by 196 countries at the 21st Conference of Parties (COP 21) in Paris, on 12 December 2015 and entered into force on 4 November 2016. It aims to limit global warming to well below 2, preferably to 1.5 °C, compared to pre-industrial levels. For further details, see: https://unfccc.int/process-and-meetings/the-paris-agreement/the-paris-agreement.
2. Sustainability or ESG ratings are scores assessing entities on criteria related to ESG themes of relevance for business or society (or both). The concept

covers a wide variety of ratings, as at present there is no legal definition of what constitutes an ESG rating. In practice, these indicators show a limited comparability, due to differences in measurement of variables and aggregation methodologies. Adding to this problem is the lack of transparency of methodologies. A third problem is the difference between relative ratings (positioning entities among their sector peers) and absolute ratings (comparing all rated entities on the same scale) and the fact that this distinction is not always clear (for example, oil companies can receive higher ratings than firms active in cleaner industries; see also Chapter 5).

3. The major providers (based on company, country and market coverage) are between 10 and 15 and include Bloomberg, CDP, FTSE Russell, ISS-ESG, MSCI, Refinitiv, RepRisk, RobecoSAM, Sustainalytics and Vigeo Eiris (European Commission, 2020).
4. Company may both disclose and buy data from sustainability-related product and service providers. According to Foubert (2020), firms mainly use ESG ratings to benchmark themselves across their industry peer group.
5. The six Principles for Responsible Investment are a voluntary set of principles suggesting possible actions for incorporating ESG issues into investment practice. They have been developed by a UN-supported network of investors to promote sustainable investment (https://www.unpri.org/signatories/reporting-and-assessment/public-signatory-reports).
6. Among the main providers, who have mostly developed in-house data services, there are FTSE Russell, MSCI, S&P and STOXX (European Commission, 2020).
7. Additional stakeholders include organisations offering advisory services or supporting companies in measurement, benchmarking, reporting aspects of their sustainability-related performance or helping investors in the definition of sustainability-related investment strategies. A few well-known examples include Ceres, Science Based Targets, Principles for Responsible Investment, World Business Council for Sustainable Development (WBCSD) and World Benchmarking Alliance (European Commission, 2020).
8. According to Tschopp and Nastanski (2014), the measurement issue is more pertinent for CSR because financial reporting has a long history and is largely standardised while CSR reporting is a more recent development where little standardisation has been achieved so far.
9. Authors develop a data-set measuring firm investments on material sustainability issues consistent with the guidance on materiality from the Sustainability Accounting Standards Board (SASB). SASB uses the SEC definition of materiality as interpreted by the US Supreme Court as the Public Company Accounting Oversight Board (PCAOB) does in its materiality guidance. Like the PCAOB, SASB defines material information

as information that represents a substantial likelihood that its disclosure will be viewed by the reasonable investor as significantly altering the total mix of information made available [SASB 'is an independent 501(c)3 non-profit whose mission is to develop and disseminate sustainability accounting standards that help publicly-listed corporations disclose material factors in compliance with SEC requirements. SASB standards are designed for the disclosure of material sustainability issues in mandatory SEC filings, such as the Form 10-K and 20-F. SASB is accredited to establish sustainability accounting standards by the American National Standards Institute (ANSI), and such accreditation is intended to signify that SASB's procedures to develop sustainability accounting standards meet the Institute's requirements for openness, balance, consensus and due process'].

10. In addition to the risks detailed in the text (i.e. physical risk and transition risk), some authors identify legal or liability risk as a separate category. The risk is specific to individual institutions active in the liability insurance market rather than to the system as a whole (Batten et al., 2016).
11. Based also on scenario analyses and geospatial analyses developed so far across regions and sectors and over time, the socioeconomic impact of climate hazards can be grouped into five categories (McKinsey Global Institute, 2020). First, heatwaves and/or permanent climate warming can put human lives at risk or at least affect human health and worsen the capability to work outdoors. Food production could also be disrupted in many areas, although in some regions it might record a benefit. Extreme events, such as precipitation or tidal flooding, can damage physical assets, like buildings, as well as service infrastructures. Finally, climate changes are destroying many forms of natural capital that in turn jeopardises the human habitat and economic activity.
12. The framework of analysis gets further complicated if redistributive aspects of adaptation are factored in, such as physical risk shifting from one group to another group of economic agents or certain losses becoming uninsurable. This might be the case for losses on properties in areas vulnerable to floods, fires or hurricanes, while households and non-financial companies, as well as governments acting as ultimate insurers of last resort would bear higher costs (Giuzio et al., 2019).
13. Internal carbon pricing includes shadow carbon pricing and virtual carbon pricing. The former is a carbon value (determined by the company) that is incorporated into investment decisions and assigned to the greenhouse gas emissions generated by projects; the latter is a levy voluntarily applied by firms to their operations that raises operating costs depending on the associated greenhouse gas emissions (CDP, 2021).
14. Based on over worldwide 5900 firms disclosing data, CDP (2021) reports that in 2020 the median price of a metric ton of CO_2 emissions ranged

at the industrial level from 16 US$ in hospitality sector to 82 US$ in apparel, while at the geographic area it ranged from 8 US$ in Asia and Latin America to 28 US$ in Europe. As of 2019, existing carbon pricing schemes are estimated to cover only about 20% of global emissions (Rosenbloom et al., 2020).

15. A flavour of the magnitude of the disruptive potential of the legislation contrasting climate changes can be captured with respect to the Paris Agreement, whose well-known goal is containing global warming in this century below 2 °C rise compared to pre-industrial levels. McGlade and Ekins (2015) show among other things that to meet this objective, at the global level, a third of oil reserves, half of gas reserves and over 80% of current coal reserves should remain unused till 2055. This would heavily impact on fossil fuel-dependent sectors both directly and indirectly as (Cahen-Fourot et al., 2019).
16. A flavour of the magnitude of the disruptive potential of the legislation contrasting climate changes can be captured with respect to the Paris Agreement, whose well-known goal is containing global warming in this century below 2 °C rise compared to pre-industrial levels. McGlade and Ekins (2015) show among other things that to meet this objective, at the global level, a third of oil reserves, half of gas reserves and over 80% of current coal reserves should remain unused till 2055.
17. Ansari and Holz (2020) investigate the potential for asset stranding and green transition in Middle East, China and Latin America over four energy and climate scenarios and find that the overall risk is moderate on average, although the burden sharing across regions is heterogeneous and some sectors (crude oil in the Middle Eastern and Latin American coal in Chine) are hit more than others thus calling for special attention.
18. See, for example, Morris et al. (2020), on the implications of the transition to a carbon-constrained economy on three coal-reliant county governments in the US.
19. For example, corporate collection and reporting of carbon footprints of economic businesses are scarce and at their first steps. The state of play is even worse when considering potential future exposure to extreme climate events. This explain why climate stress test exercises are complex and mainly focus on the near-term impact of climate policies.
20. In the absence of adaptive measures, the costs related to physical risk steadily increase, thus raising both the magnitude of responses needed in the future and the likelihood of a too late, too sudden scenario where effective mitigation of physical risks is no longer feasible (Carney, 2015; ESRB, 2016; Lane, 2019; Vermeulen et al., 2018; Weyzig et al., 2014). In contrast, early actions would lessen physical risk and allow a gradual repricing of assets, while entailing both significant short-term upfront costs compensated by potential long-term positive benefits. According to

available estimates, the investments needed for the transition to a low-carbon economy range between USD 1 trillion and USD 4 trillion in constant terms when considering the energy sector alone, or up to USD 20 trillion when encompassing a broader economic spectrum. At the same time, these estimates neglect the embedded benefits in terms of employment or output multipliers of such investments, as well as productivity gains associated with new technologies (ECB & ESRB, 2020).

21. The Nomenclature statistique des activités économiques dans la Communauté européenne (NACE) is the Statistical classification of economic activities in the European Community (for further details see https://ec.europa.eu/eurostat/statistics-explained/index.php?title=Glossary:Statistical_classification_of_economic_activities_in_the_European_Community_(NACE)).
22. Based on the ECB's Securities Holdings Statistics by Sector, Giuzio et al. (2019) find that in recent years investment and pension funds have reduced their relative exposures to securities issued by climate policy-relevant sectors (in some cases probably following policy guidelines on asset allocation), while banks and insurance firms kept their exposures relatively constant. Battiston et al. (2017) remap all the sectors at NACE2 4-digit level into new climate policy-sensitive sectors, combining criteria including carbon emissions, the role of the sector in the supply chain, and the existence of traditional policy institutions for the sectors.
23. The stress test encompassed about four million companies worldwide and 2000 banks—almost all in the euro area—and covered a 30-year time horizon in the future. The objective was to test the resilience of banks and non-financial firms in a range of three scenarios: the 'orderly transition' scenario, based on the timely implementation of effective climate policies; on the other extreme the 'hot house world' scenario, assuming no new climate policies; the 'disorderly transition' scenario, considering the effect of a delayed and abrupt implementation of policy actions. This stress test exercise is a 'top-down' exercise, run centrally by the ECB staff and based on internal datasets and models. It differs from the 'bottom-up' supervisory climate stress test, to be held in 2022, which instead will use banks' self-assessed data on their exposure to climate change risk and the measures put in place to address it.
24. One of the main challenge is gauging climate change risk exposure at the individual firm level as well as its evolution over-time, given the great heterogeneity among firms and the uncertainty on the impact of a changing climate across sectors (Xu et al., 2021). An additional point is the need for disentangling the physical and the transitional dimensions of climate risk as they may trigger opposite market reactions: for example, enhanced curbing of carbon emissions driven by events like the Paris agreement can raise transition risk while lowering physical risk.

25. There is evidence that this premium has been increasing since the Paris agreement to reduce worldwide emissions; see P. Bolton and M. Kacperzyk (2021), Global Pricing of Carbon-Transition Risk, Working Paper No. w28510, National Bureau of Economic Research.
26. For example, Mésonnier and Nguyen (2021) find that mandatory disclosures for French institutional investors led to divestment from fossil fuel companies. A recent report from the NGFS (2021) also highlighted that investors in certain energy-intensive sectors may be more sensitive to climate disclosures by issuers. Alessi et al. (2021) show the existence of a negative 'greenium' in the European equity market.
27. High Frequency Trading (HFT) is a type of algorithmic trading that allows to buy and sell large volumes of securities automatically at a very high speed (Caivano et al., 2012).
28. Among the studies pointing to a negative impact of the HFT on the informational efficiency of financial markets and price discovery see Jarrow and Protter (2012) and Zhang (2010).
29. Among the studies pointing to a positive impact of the HFT on market liquidity see Menkveld (2013), Hasbrouck and Saar (2013) and Riordan and Storkenmaier (2011).
30. The EMH rests on the assumptions that rational market players, with properly identified subjective utility functions, are capable to rationally compute any asset fundamental value as the net present value of its prospect cash flows discounted by a risk factor. In addition, the EMH posits that the trades of irrational individuals either cancel out with one another or are arbitraged away by rational investors (Fama, 1965).

References

2 Degrees Investing. (2017). *All swans are black in the dark.* https://2degrees-investing.org/wp-content/uploads/2017/02/All-swans-are-black-in-the-dark.pdf

Albouy, D., Graf, W., Kellogg, R., & Wolff, H. (2013). *Climate amenities, climate change, and American quality of life* (NBER Working Paper, No. 18925). https://www.nber.org/system/files/working_papers/w18925/w18925.pdf

Alessi, L., Ossola, E., & Panzica, R. (2021). What greenium matters in the stock market? The role of greenhouse gas emissions and environmental disclosures. *Journal of Financial Stability, 54*(C).

Amzallag, A. (2021). Fund portfolio networks: A climate risk perspective. *ESMA trends risks vulnerabilities 2.* https://www.esma.europa.eu/sites/default/files/library/esma50-165-1842_trv2-2021.pdf

Ansari, D., & Holz, F. (2020). Between stranded assets and green transformation: Fossil-fuel-producing developing countries towards 2055. *World Development*, 130. https://doi.org/10.1016/j.worlddev.2020.104947

Atanasova, C., & Schwartz, E. S. (2019). *Stranded fossil fuel reserves and firm value* (NBER Working Paper, No. 6497). https://www.nber.org/papers/w26497

Bansal, R., Kiku, D., & Ochoa, M. (2016). *What do capital markets tell us about climate change?* (Working Paper). Duke University. https://economicdynamics.org/meetpapers/2016/paper_542.pdf

Barton, D. (2017). Refocusing capitalism on the long term: Ownership and trust across the investment value chain. *Oxford Review of Economic Policy, 33*(2), 188–200.

Batten, S. (2018). *Climate change and the macro-economy: A critical review* (Bank of England Staff Working Paper, 706). https://www.bankofengland.co.uk/working-paper/2018/climate-change-and-the-macro-economy-a-critical-review

Batten, S., Sowerbutts, R., & Tanaka, M. (2016). *Let's talk about the weather: The impact of climate change on central banks* (Bank of England Staff Working Paper, 603). https://www.bankofengland.co.uk/working-paper/2016/lets-talk-about-the-weather-the-impact-of-climate-change-on-central-banks

Battiston, S., Mandel, A., Monasterolo, I., Schütze, F., & Visentin, G. (2017). A climate stress-test of the financial system. *Nature Climate Change, 7*(4), 283–288.

Bauer, R., & Hann, D. (2010). *Corporate environmental management and credit risk* (ECCE Working Paper). University Maastricht, The European Centre for Corporate Engagement. https://citeseerx.ist.psu.edu/viewdoc/download?doi=10.1.1.800.173&rep=rep1&type=pdf

Board of Governors of the Federal Reserve System. (2020, November). *Financial Stability Report*. https://www.federalreserve.gov/publications/files/financial-stability-report-20201109.pdf

Bolton, P., & Kacperczyk, M. T. (2021). Do investors care about carbon risk? *Journal of Financial Economics*. https://www.sciencedirect.com/science/article/abs/pii/S0304405X21001902?via%3Dihub

Bowen, A., & Dietz, S. (2016). *The effects of climate change on financial stability, with particular reference to Sweden, A report for Finansinspektionen*. https://www.lse.ac.uk/business/consulting/reports/the-effects-of-climate-change-on-financial-stability-with-particular-reference-to-sweden

Bua, G., Kapp, D., Ramella, F., & Rognone, L. (2021). *Transition versus physical climate risk pricing in euro area financial markets: A text-based approach*. https://papers.ssrn.com/sol3/papers.cfm?abstract_id=3860234

Bunten, D., & Kahn, M. E. (2014). *The impact of emerging climate risks on urban real estate price dynamics* (NBER Working Paper Series, 20018). Cambridge MA

Cahen-Fourot, L., Campiglio, E., Dawkins, E., Godin, A., & Kemp-Benedict, E. (2019). *Capital stranding cascades: The impact of decarbonisation on productive asset utilisation* (WU Working Paper Series, 18/2019). https://epub.wu.ac.at/8007/

Caivano, V., Ciccarelli, S., Di Stefano, G., Fratini, M., Gasparri, G., Giliberti, M., Linciano, N., & Tarola, I. (2012). *High frequency trading: Definition, effects, policy issues* (CONSOB Discussion Paper, 5). http://www.consob.it/documents/46180/46181/dp5en.pdf/21c2da8d-e815-49ce-ad18-b183f12ffaa4

Campagnolo, L., & Davide, M. (2019). Can the Paris deal boost SDGs achievement? An assessment of climate mitigation co-benefits or side-effects on poverty and inequality. *World Development, 122*, 96–109. https://doi.org/10.1016/j.worlddev.2019.05.015

Capelle-Blancard, G., Crifo, P., Diaye, M. A., Oueghlissi, R., & Scholtens, B. (2017). *Environmental, Social and Governance (ESG) performance and sovereign bond spreads: An empirical analysis of OECD countries* (University of Paris Nanterre EconomiX Working Papers, 2017-7).

Carney, M. (2015, September 29). *Breaking the tragedy of the horizon—Climate change and financial stability, speech, Lloyd's of London*. https://www.bankofengland.co.uk/speech/2015/breaking-the-tragedy-of-the-horizon-climate-change-and-financial-stability

Carney, M. (2018, April 6). *A transition in thinking and action: Remarks at the international climate risk conference for supervisors*. De Nederlandsche Bank. https://www.bis.org/review/r180420b.pdf

CDP. (2021). *Putting a price on carbon: The state of internal carbon pricing by corporates globally*. https://6fefcbb86e61af1b2fc4c70d8ead6ced550b4d987d7c03fcdd1d.ssl.cf3.rackcdn.com/cms/reports/documents/000/005/651/original/CDP_Global_Carbon_Price_report_2021.pdf?1618938446

Cevik, S., & Jalles, J. (2020). *Feeling the heat: Climate shocks and credit ratings* (International Monetary Fund Working Paper 2020/286).

CFTC. (2020). *Managing climate risk in the U.S. financial system*. Report of the Climate-Related Market Risk Subcommittee, Market Risk Advisory Committee of the U.S. Commodity Futures Trading Commission. https://www.cftc.gov/PressRoom/PressReleases/8234-20

Choi, D., Gao, Z., & Jiang, W. (2020). Attention to global warming. *The Review of Financial Studies, 33*(3), 1112–1145. https://doi.org/10.1093/rfs/hhz086

Ciscar, J. C., Feyen, L., Ibarreta, D., & Soria, A. (Eds.). (2018). *Climate impacts in Europe: Final report of the JRC PESETA III project*. Publications Office of the European Union.

Clark, G. L., Feiner, A., & Viehs, M. (2015). *From the stockholder to the stakeholder. How sustainability can drive financial outperformance*. University of Oxford and Arabesque Partners. https://arabesque.com/research/From_the_stockholder_to_the_stakeholder_web.pdf

Clark, G. L., & Viehs, M. (2014). *The implication of corporate social responsibility for investors: An overview and evaluation of the existing CSR literature* (European Centre for Corporate Engagement Working Paper). https://sustainable-finance.nl/research/the-implications-of-corporate-social-responsibility-for-investors/

Cremers, K. J. M., & Nair, V. B. (2005). Governance mechanisms and equity prices. *Journal of Finance, 60*(6), 2859–2894. https://sustainable-finance.nl/research/the-implications-of-corporate-social-responsibility-for-investors/

Cremers, K. J. M., Nair, V. B., & Wei, C. (2007). Governance mechanisms and bond prices. *Review of Financial Studies, 20*(5), 1359–1388. https://econpapers.repec.org/article/ouprfinst/v_3a20_3ay_3a2007_3ai_3a5_3ap_3a1359-1388.htm

DeLisle, J., French, D., & Vivero, M. G. (2015). *Comovement, passive investing and price informativeness*. SPGMI: Compustat Fundamentals (Topic).

Deutsches Aktieninstitut & Rothschild & Co Deutschland. (2018). *ESG from the perspective of institutional investors*. https://www.rothschildandco.com/siteassets/publications/rothschildandco/global_advisory/2019/en_ga_2019_esg_from_the_perspective_of_institutional_investors.pdf

ECB. (2021). *Shining a light on climate risks: The ECB's economy-wide climate stress test*. https://www.ecb.europa.eu/press/blog/date/2021/html/ecb.blog210318~3bbc68ffc5.en.html

ECB & ESRB. (2020). *Positively green: Measuring climate change risks to financial stability*. https://www.esrb.europa.eu/pub/pdf/reports/esrb.report200608_on_Positively_green_-_Measuring_climate_change_risks_to_financial_stability~d903a83690.en.pdf?c5d033aa3c648ca0623f5a2306931e26

Eccles, R. G., Ioannou, I., & Serafeim, G. (2014). The impact of corporate sustainability on organizational processes and performance. *Management Science, 60*(11), 2835–2857. http://dx.doi.org/10.2139/ssrn.1964011

Eccles, R. G., & Serafeim, G. (2013). The performance frontier: Innovating for a sustainable strategy. *Harvard Business Review*, 50–60. https://hbr.org/2013/05/the-performance-frontier-innovating-for-a-sustainable-strategy

Engle, R. F., Giglio, S., Kelly, B., Lee, H., & Stroebel, J. (2020). Hedging climate change news. *The Review of Financial Studies, 33*(3), 1184–1216. https://doi.org/10.1093/rfs/hhz072

ESMA. (2020). *Trends risks vulnerabilities 2*. https://www.esma.europa.eu/sites/default/files/library/esma_50-165-1287_report_on_trends_risks_and_vulnerabilities_no.2_2020.pdf

ESMA. (2021). *Trends risks vulnerabilities 2*. https://www.esma.europa.eu/sites/default/files/library/esma50-165-1842_trv2-2021.pdf

ESRB. (2016). *Too late, too sudden: Transition to a low-carbon economy and systemic risk*. ESRB Reports of the Advisory Scientific Committee, 6. https://www.esrb.europa.eu/pub/html/index.en.html

European Commission. (2020). *Study on sustainability-related ratings, data and research*. https://op.europa.eu/en/publication-detail/-/publication/d7d85036-509c-11eb-b59f-01aa75ed71a1/language-en/format-PDF/source-183474104

Fama, E. F. (1965). Random walks in stock market prices. *Financial Analysts Journal, 21*, 55–59. https://www.jstor.org/stable/4479810

Fama, E. F. (1998). Market efficiency, long-term returns, and behavioral finance. *Journal of Financial Economics, 49*(3), 283–306.

Foubert, A. L. (2020). ESG data market: No stopping its rise now. *Opimas*. http://www.opimas.com/research/547/detail/

Friede, G., Busch, T., & Bassen, A. (2015). ESG and financial performance: Aggregated evidence from more than 2000 empirical studies. *Journal of Sustainable Finance & Investment, 5*(4), 210–233. https://www.tandfonline.com/doi/full/10.1080/20430795.2015.1118917

FSB. (2021). *The availability of data with which to monitor and assess climate-related risks to financial stability*. https://www.fsb.org/wp-content/uploads/P070721-3.pdf

Galant, A., & Cadez, S. (2017). Corporate social responsibility and financial performance relationship: A review of measurement approaches. *Economic Research- Ekonomska Istraživanja, 30*(1), 676–693. https://www.tandfonline.com/doi/full/10.1080/1331677X.2017.1313122

Gassebner, M., Keck, A., & Teb, R. (2010). Shaken, not stirred: The impact of disasters on international trade. *Review of International Economics, 18*(2), 351–368 https://onlinelibrary.wiley.com/doi/abs/10.1111/j.1467-9396.2010.00868.x

Giuzio, M., Krusec, D., Levels, A., Melo, A. S., Mikkonen, K., & Radulova, P. (2019). Climate change and financial stability. *Financial Stability Review*. European Central Bank. https://www.ecb.europa.eu/pub/financial-stability/fsr/special/html/ecb.fsrart201905_1~47cf778cc1.en.html

Greening, D. W., & Turban, D. B. (2000). Corporate social performance as a competitive advantage in attracting a quality workforce. *Business & Society, 39*(3), 254–280. https://journals.sagepub.com/doi/10.1177/000765030003900302

Hallegatte, S. (2009). Strategies to adapt to an uncertain climate change. *Global Environmental Change, 19*(2), 240–247. https://doi.org/10.1016/j.gloenvcha.2008.12.003

Hasbrouck, J., & Saar, G. (2013). *Low-latency tradin* (Johnson School Research Paper Series, 35–2010). AFA 2012 Chicago Meetings Paper. http://ssrn.com/abstract=1695460 or http://dx.doi.org/10.2139/ssrn.1695460

Hong, H., Li, F., & Xu, J. (2018). Climate risks and market efficiency. *Journal of Econometrics, 208*, 265–281. https://doi.org/10.1016/j.jeconom.2018.09.015

IMF. (2020, October). *Global Financial Stability Report: Bridge to Recovery.*

Investment Leaders Group, University of Cambridge Institute for Sustainability Leadership (ILG). (2014). *The value of responsible investment*. https://www.cisl.cam.ac.uk/resources/sustainable-finance-publications/value-responsible-investment

Jarrow, R. A., & Protter, P. (2012). A dysfunctional role of high frequency trading in electronic markets. *International Journal of Theoretical and Applied Finance, 15*(3). https://doi.org/10.1142/S0219024912500227

Kahneman, D. (2013). *Thinking, fast and slow*. Farrar, Straus and Giroux publisher.

Kenechukwu, A., Kruttli, M., McCabe, P., & Osambela, E. (2020). *The shift from active to passive investing: Potential risks to financial stability?* (Finance and Economics Discussion Series 2018–060r1). Board of Governors of the Federal Reserve System. https://doi.org/10.17016/FEDS.2018.060r1

Khan, M., Serafeim, G., & Yoon, A. S. (2016). Corporate sustainability: First evidence on materiality. *The Accounting Review, 91*(6), 1697–1724. https://doi.org/10.2139/ssrn.2575912

KPMG. (2014). *A new vision of value: Connecting corporate and societal value creation*. https://home.kpmg/xx/en/home/insights/2014/09/a-new-vision-connecting-corporate.html

Kruger, P. (2015). Corporate goodness and shareholder wealth. *The Journal of Finance Economics, 115*(2), 304–329. http://www.sciencedirect.com/science/article/pii/S0304405X14001925

Kumar, A., Xin, W., & Zang, C. (2019). *Climate sensitivity and predictable returns*. http://dx.doi.org/10.2139/ssrn.3331872

Kurucz, E. C., Colbert, B. A., & Wheeler, D. (2008). The business case for corporate social responsibility. In A. Crane, A. McWillimas, D. Matten, J. Moon, & D. S. Siegel (Eds.), *The Oxford handbook of corporate social responsibility*. Oxford University Press. https://doi.org/10.1093/oxfordhb/9780199211593.003.0004

Lane, P. R. (2019). *Climate change and the Irish financial system*. Central Bank of Ireland. https://www.centralbank.ie/docs/default-source/publications/economic-letters/vol-2019-no-1-climate-change-and-the-irish-financial-system-(lane).pdf

Lo, A. W. (2017). *Adaptive markets: Financial evolution at the speed of thought*. Princeton University Press. https://press.princeton.edu/books/paperback/9780691191362/adaptive-markets

Mazzacurati, J. (2021). *ESG ratings: Status and key issues ahead* (ESMA TRV No. 1). https://www.esma.europa.eu/sites/default/files/library/esma50-165-1524_trv_1_2021.pdf

McGlade, C., & Ekins, P. (2015). The geographical distribution of fossil fuels unused when limiting global warming to 2 °C. *Nature, 517*, 187–190. https://doi.org/10.1038/nature14016

McKinsey Global Institute. (2020). *Climate risk and response: Physical hazards and socioeconomic impacts*. https://www.mckinsey.com/business-functions/sustainability/our-insights/climate-risk-and-response-physical-hazards-and-socioeconomic-impacts

Mésonnier, J.-S., & Nguyen, B. (2021). *Showing off cleaner hands: Mandatory climate-related disclosure by financial institutions and the financing of fossil energy*. (Working paper no. 800, Banque de France).

Menkveld, A. J. (2013). High frequency trading and the new-market makers. *Journal of Financial Markets, 16*. http://ssrn.com/abstract=1722924 or https://doi.org/10.2139/ssrn.1722924

Morris, A. C., Kaufman, N., & Doshi, S. (2020). Revenue at risk in coal-reliant counties. In M. Kotchen, J. H. Stock, & C. Wolfram (Eds.), *Environmental and energy policy and the economy* (p. 2). University of Chicago Press. https://doi.org/10.1086/711307

Nagy, Z., Kassam, A., & Lee, L. E. (2016). Can ESG add alpha? An analysis of ESG tilt and momentum strategies. *Journal of Investing, 25*(2), 113–124. https://doi.org/10.3905/joi.2016.25.2.113

NGFS (Network for Greening the Financial System). (2019). *First comprehensive report: A call for action. Climate change as a source of financial risk*. https://www.ngfs.net/sites/default/files/medias/documents/ngfs_first_comprehensive_report_-_17042019_0.pdf

NGFS (Network for Greening the Financial System). (2020a, June). *NGFS climate scenarios for central banks and supervisors*.

NGFS (Network for Greening the Financial System). (2020b). *Integrating climate related and environmental risks in prudential supervision—Guide for Supervisors*. Network for Greening the Financial System.

NN Investment Partners and ECCE (European Centre for Corporate Engagement). (2017). *The materiality of ESG factors for emerging markets equity investment decisions: Academic evidence*. https://yoursri.be/media-new/download/nn_ip_-_ecce_emerging_markets_equities.pdf

OECD. (2019). *Aligning development co-operation and climate action: The only way forward, the development dimension*. OECD Publishing. https://doi.org/10.1787/5099ad91-en

OECD. (2021). *Financial markets and climate transition: Opportunities, challenges and policy implications*. https://www.oecd.org/finance/financial-markets-and-climate-transition-opportunities-challenges-and-policy-implications.htm

Opitz Stapleton, S., Nadin, R., Watson, C., & Ketter, J. (2017). *Climate change, migration and displacement: The need for risk informed and coherence approach*. Overseas Development Institute.

Ortiz-de-Mandojana, N., & Bansal, P. (2016). The long-term benefits of organizational resilience through sustainable business practices. *Strategic Management Journal, 37*(8), 1615–1631. https://doi.org/10.1002/smj.2410

Pastor, L., Stambaugh, R. F., & Taylor, L. A. (2019). *Sustainable investing in equilibrium* (NBER Working Paper Series, 26549). https://ideas.repec.org/p/nbr/nberwo/26549.html

Polbennikov, S., Desclée, A., Dynkin, L., & Maitra, A. (2016). ESG ratings and performance of corporate bonds. *The Journal of Fixed Income, 26*(1), 21–41. https://www.econbiz.de/Record/esg-ratings-and-performance-of-corporate-bonds-polbennikov-simon/10011660749

Porter, M. E., & Kramer, M. R. (2006, December). Strategy and society: The link between competitive advantage and corporate social responsibility. *Harvard Business Review*, 78–93. https://hbr.org/2006/12/strategy-and-society-the-link-between-competitive-advantage-and-corporate-social-responsibility

PRI-UN Global Compact. (2013). *The value-driver Model: A tool for communicating the business value of sustainability*. https://www.unglobalcompact.org/take-action/action/value-driver-model

Riordan, R., & Storkenmaier, A. (2011). Latency, liquidity and price discovery. *Journal of Financial Markets, 15*(4). https://doi.org/10.2139/ssrn.1247482

Roberts, P. W., & Dowling, G. R. (2002). Corporate reputation and sustained superior financial performance. *Strategic Management Journal, 23*, 1077–1093. https://doi.org/10.1002/smj.274

Rosenbloom, D., Markard, J., Geels, F. W., & Fuenfschilling, L. (2020). Why carbon pricing is not sufficient to mitigate climate change—And how "sustainability transition policy" can help. *Proceedings of the National Academy of Science of the United States of America, 117*(16), 8664–8668. https://doi.org/10.1073/pnas.2004093117

S&P Global. (2019). *Understanding climate risk at the asset level: The interplay of transition and physical risks*. https://www.spglobal.com/en/research-insights/featured/understanding-climate-risk-at-the-asset-level-the-interplay-of-transition-and-physical-risks

Schoenmaker, D., & Schramade, W. (2019). *Principles of sustainable finance*. Oxford University Press.

Schoenmaker, D., & Schramade, W. (2020). *Asset pricing and sustainability: A teaching note*. https://ssrn.com/abstract=3479730

Shashwat, A., Kumar, N., & Wermers, R. R. (2020). Do fund managers misestimate climatic disaster risk? *Review of Financial Studies, Society for Financial Studies, 33*(3), 1146–1183. http://hdl.handle.net/10.1093/rfs/hhz143

Stein, J. C. (1989). Efficient capital markets, inefficient firms: A model of myopic corporate behaviour. *Quarterly Journal of Economics, 104*(4), 655–669. https://doi.org/10.2307/2937861

Tschopp, D., & Nastanski, M. (2014). The harmonization and convergence of corporate social responsibility reporting standards. *Journal of Business Ethics, 125*, 147–162. http://hdl.handle.net/10.1007/s10551-013-1906-9

UN Global Compact. (2017). *United Nations Global Compact progress report: Business solutions to sustainable development*. UN Global Compact. https://www.unglobalcompact.org/library/5431

Vermeulen, R., Schets, E., Lohuis, M., Kölbl, B., Jansen, D. J., & Heeringa, W. (2018). An energy transition risk stress test for the financial system of the Netherlands. *Occasional Studies, 16*(7). De Nederlandsche Bank (DNB). https://www.dnb.nl/en/publications/research-publications/occasional-studies/nr-7-2018-an-energy-transition-risk-stress-test-for-the-financial-system-of-the-netherlands/

Weyzig, F., Kupper, B., van Gelder, J. W., & van Tilburg, R. (2014, February). The price of doing too little too late—The impact of the carbon bubble on the EU financial system. Report prepared for the Greens/EFA Group, European Parliament, 2014. https://gef.eu/publication/carbon-bubble-price-little-late/

World Bank. (2021). *State and trends of carbon pricing*. https://openknowledge.worldbank.org/handle/10986/35620

Wu, M. W., & Shen, C. H. (2013). Corporate Social responsibility in the banking industry: Motives and financial performance. *Journal of Banking and Finance, 37*(9), 3529–3547. http://www.sciencedirect.com/science/article/pii/S0378426613002069

Xu, J., Sun, C., & You, Y. (2021). *Climate change exposure and stock return predictability*. https://papers.ssrn.com/sol3/papers.cfm?abstract_id=3777060

Zhang, F. (2010). *High-frequency trading, stock volatility and price discovery*. http://ssrn.com/abstract=1691679

PART II

The Information as a Trigger: Market Failures and Regulatory Issues

CHAPTER 4

Corporate Disclosure

Cristiano Busco, Alessandro D'Eri, and Valerio Novembre

Introduction

Whether large, medium, or small, every year companies invest resources, time, and energy as their intertwined cycles of planning, management, measurement, and reporting unfold recursively in practice. Designed and prepared to fulfil the expectations of external and internal stakeholders,

The opinions stated by the authors are personal and do not bind ESMA.

C. Busco
LUISS University, Rome, Italy
e-mail: cbusco@luiss.it

A. D'Eri (✉)
European Securities and Markets Authority, Paris, France
e-mail: alessandro.deri@esma.europa.eu

V. Novembre
European Securities and Markets Authority, Paris, France
e-mail: valerio.novembre@esma.europa.eu

N. Linciano et al. (eds.), *Information as a Driver of Sustainable Finance*, Palgrave Studies in Impact Finance,
https://doi.org/10.1007/978-3-030-93768-3_4

amongst which the current and potential shareholders, corporate disclosures are positioned at the final stage of these cycles as they highlight and summarize the processes of strategic planning, operations management, and performance measurement that are in place. Within this context, and in the attempt to offer an account of the ongoing search for competitiveness and sustainable growth featuring contemporary organizations across industries, corporate reports are nowadays expected to convey insightful information beyond the traditional key financial data. Arguably, these reports are increasingly considered to be worthwhile reading due to their ability to enlighten readers about the way in which corporate purpose drive companies' strategies and business models to align with the opportunities offered by the market in the attempt to generate sustainable performance.

However, the concepts, elements, and principles that characterize the way in which organizations plan, manage, measure, and report their annual performance, as well as the relevance of corporate disclosures, have all been questioned over the last two decades regarding several aspects, ranging from the supposed compliance-driven content they provide to the failure of offering forward-looking information about company strategy, performance, and risk. This perception has worsened in the aftermath of the recent financial crises (most recently, that of 2008 and stretching up until 2011) and corporate scandals, in which several commentators and analysts have firmly portrayed the global economic system as broken and have viewed organizations as being one of the major causes of social, environmental, and economic problems. This has led many experts, as well as public opinion, to criticize these fundamental documents as gradually becoming less fit for the purpose and often failing to shed light on strategy execution as well as on the extensive variety of intangible and non-financial resources used within an organization's processes of value creation.

This was the situation, and its complications, at the beginning of 2020, when Covid-19 hit the world to make expectations for change urgent and higher. Thus, in an era where the impacts of a global pandemic, climate change and growing inequality are intensifying, intangible value and the need for data-driven sustainability information have sharply grown in importance. Capital markets, as well as Society as a whole, demand evidence-based, market-informed, and transparent data to deliver long-term value to shareholders while also helping secure the future of our people and our planet. That is why within this chapter we will focus on corporate disclosure as a fundamental driver of information to make Sustainable Finance happen.

First, we explore the corporate disclosure landscape as non-financial information becomes paramount. We begin emphasizing the importance for contemporary organization to communicate the building blocks of their licence to operate, namely: corporate purpose, sustainable strategy and business models, material issues and risks, innovative products and services, as well as their impact on the multitude of heterogeneous stakeholders and society at large. In addition, we will briefly summarize the recent evolutions in the unfolding scenario of non-financial reporting and Environmental, Social and Governance (ESG) disclosure, whose actors seem to work towards a certain degree of simplification and consolidation.

Then, we reflect on two important recent developments that have impacted on the European corporate disclosure landscape: the Non-Financial Reporting Directive and the revised Shareholder Rights Directive. The Non-Financial Reporting Directive will review and discuss a series of aspects, such as the scoping requirements of the Directive; the minimum compliance by reporting entities with the basic requirements; the broader context surrounding the disclosures; the notion of materiality applied; the extent to which the level of disclosure has been regarded as adequate to meeting the information needs of various stakeholders; and, finally, the areas for future improvement of the regime for non-financial reporting. Regarding the revised Shareholder Rights Directive, we discuss the new provisions—in the area of directors' remuneration—that have the highest impact on corporate disclosures related to ESG factors.

The Corporate Disclosure Landscape as Non-financial Information Becomes Paramount

Companies' Necessity to Communicate the Building Blocks of Their Licence to Operate

Contemporary organizations operate in a complex and uncertain world, characterized by many internal and external drivers, interdependencies, and trade-offs that influence the process of decision-making, the promises that these decisions entail, and the expectations of a variety of demanding stakeholders. Organizational leaders are increasingly required to navigate these challenges by implementing a comprehensive approach to strategic planning, performance measurement, and reporting. Ultimately, if they intend to maintain and reinforce their "license to operate", companies are increasingly required to ally corporate *purpose* with *strategy*, as well as to transform their purposeful strategies eventually presented to markets and investors in product, service, and process innovations able to deliver and communicate *impact*—i.e. achieve the results expected by a multitude of heterogeneous stakeholders and share key information with them.

Formulating and articulating corporate purpose is likely to be challenging, in some cases even uncomfortable, since it may require going beyond a mere description of the company's "business-as-usual" to reflect opportunities for impactful business innovation and disruption. It often requires rethinking the organization and its strategy as it develops at the intersection of aspiration, inclusion, and action.[1] Corporate purpose should aim to articulate a long-term inclusive aspiration that a company is willing and able to act on, using existing resources and relationships or capabilities it could develop. According to Larry Fink, CEO of BlackRock, "Society is demanding that companies, both public and private, serve a social purpose. To prosper over time, every company must not only deliver financial performance, but also show how it makes a positive contribution to society".[2] As Fink's statement suggests, an effective Purpose requires mediating, integrating, and balancing a number of tensions amongst different corporate stakeholders and their goals. And the trade-offs that result from such balancing need to be acknowledged, measured, and communicated as the organization report the efforts and performance connected to the creation of sustainable value over time.

To enable Purpose to move from "articulation" to "execution" requires connecting strategy and capital allocation with a comprehensive performance measurement system able to connect sustainability issues with financial returns.[3] It involves reconciling competitiveness and sustainable growth within the context of inclusive business models to take advantage of the opportunities and face the challenges of the market. Importantly, this requires the development of new practices of accounting and reporting that are able to capture the wholeness of processes of long-term value creation; taking account of the positive and negative social and environmental externalities that are produced by its operations and, especially, its products and services, as well as how the multiple and heterogeneous resources provided by the company's stakeholders contribute to its financial and non-financial performance. Although there continue to be laggards, the investment community now broadly recognizes the need to better understand how material ESG issues matter to financial performance.[4]

The increasing pressure from investors and growing awareness of corporate leaders have led to the development of a number of accounting and reporting frameworks that attempt to capture the relationship between non-financial (sometimes labelled as pre-financial) and financial performance. While a detailed and in-depth review of such developments is beyond the scope of this chapter, it is important to acknowledge the effort of various non-financial accounting and reporting frameworks in exploring the link between multiple performance metrics, as well as in attempting to quantify the impact of non-financial resources and material matters on the company's bottom line, as well as long-run efficiency and value. Each of the accounting and reporting frameworks has been designed to engage different audiences, ranging from investors and corporate boards to other corporate stakeholders, public bodies, and NGOs. But whatever the framework considered, understanding the concept of ESG financial materiality and impact is the place to start.

The Unfolding Scenario of Non-financial Reporting and ESG Disclosure

The first organization to provide guidance on the disclosure of a broad set of non-financial metrics was the Global Reporting Initiative (GRI).[5] Launched in 1997 the objective of the GRI was to develop and establish reporting guidelines for the "triple bottom line": accounting for economic as well as environmental and social performance by corporations. The GRI aspired to gradually evolve the so-called sustainability reporting to a point that it would be similar to financial reporting in terms of rigour, credibility, and comparability. The GRI Guidelines, whose first version was launched in 2000, included materiality guidelines to help companies determine the threshold at which sustainability issues were sufficiently important to stakeholders that they should be reported.

In contrast to the GRI's focus on reporting stand-alone sustainability information, the International Integrated Reporting Council (IIRC) was established in 2010 to integrate corporate disclosures of financial and sustainability information.[6] Integrated reporting was presented as a process founded on integrated thinking that results in a periodic integrated report by an organization about value creation over time and related communications regarding aspects of value creation. An integrated report, therefore, represents a concise communication about how an organization's strategy, governance, performance, and prospects, in the context of its external environment, lead to the creation of value in the short, medium, and long term.

The Framework released by the IIRC in December 2013 suggests that the fundamental concepts of Integrated Thinking and Reporting are represented by the capitals that an organization uses and affects, as well as the process of creating value over time. This value is embodied in the capitals—also referred to as resources and relationships. Companies depend on six different types of capitals, which are stores of value that, in one form or another, become inputs to an organization's business model. They are financial, manufactured, intellectual, human, social and relationship, and natural. The Framework does not require organizations to adopt the capitals, so they should be rather used as a benchmark to ensure an organization doesn't overlook a capital that it uses or affects. Value is created or destroyed through the capitals within a company's business model, which represents the chosen system of inputs, business activities, outputs, and outcomes that aims to create value over the short, medium, and long term.

A more recent set of voluntary standards was developed by the Sustainability Accounting Standards Board (SASB).[7] Importantly, whereas the GRI has a multi-stakeholder focus, SASB's mission was to develop and disseminate industry-specific sustainability accounting standards to help publicly listed corporations disclose "material" sustainability information to investors. With the aim of identifying the ESG issues that affect corporate financial performance and are therefore financially material for investors, SASB developed market-informed and industry-specific standards. Because the materiality of sustainability issues varies across industries, SASB has established the Sustainable Industry Classification System™ (SICS), which comprises 11 sectors that divide further into 77 industries. Within this system, companies are grouped in terms of similar resource intensity, as well as sustainability risks and opportunities.

For the purposes of developing SASB standards, sustainability as a concept has been interpreted as referring to corporate ESG activities that, in the process of addressing environmental and social issues, help to maintain or even increase the ability of the company to create value over the long term. Sustainability accounting in accordance with SASB standards refers to the measurement, management, and reporting of such corporate activities. Overall, SASB identifies 26 sustainability general issue categories (GICs) that are organized under five broad sustainability dimensions: environment, social capital, human capital, business model and innovation, and leadership and governance. For each industry-specific material issue, SASB identifies a set of disclosure topics and a subset of associated accounting metrics. Going further, for each industry, SASB has also analysed the link between disclosure topics and each of 13 financial value drivers that are grouped into the six categories of revenue, operating expenses, non-operating expenses, assets, liabilities, and cost of capital. At the industry level, each disclosure topic might impact one or more value drivers to a different extent. The greater the number of the value drivers affected, the higher is the "financial relevance" of the material issue.

The work of SASB complements that of GRI. Both initiatives are essential reporting standards that serve different ends. As such, the two frameworks are by no means to be understood as competing, but rather as "complementary" and "mutually supportive", as reported by the chief executive officers of both GRI and SASB.[8] With its focus on society at large, GRI identifies issues that, while not necessarily financially material today, can become so in the future. GRI is "the canary in the coal mine". To some extent, this will occur because stakeholders can make it so. Their actions and decisions—such as whether to buy a product or go to work for a company—can make the issue a material one for investors. The need to harmonize the several existing reporting frameworks is deemed essential.

To this end, the Corporate Reporting Dialogue (CRD) has launched the "better alignment project", which seeks to foster consistency between the different frameworks in the sustainability reporting ecosystem. The project has already yielded promising results in terms of the better alignment of the frameworks' standards to the recommendations published by the Task Force on Climate-related Financial Disclosures (TCFD). "While no framework is perfect", writes Larry Fink in his 2020 letter to CEOs, "BlackRock believes that SASB provides a clear set of standards for reporting sustainability information across a wide range of issues". And as he goes on to say, "for evaluating and reporting climate-related risks, as well as the related governance issues that are essential to managing them, the TCFD provides a valuable framework".[9] Framework alignment is important because investors today are genuinely interested in sustainability issues, but they have to sift through information that is not standardized, not comparable, and, in many cases, not material.

Towards a Simplified Scenario of ESG and Value Creation Disclosure: The IIRC-SASB Merger

On the 25th of November 2020, the IIRC and SASB announced their intention to merge into a unified organization, the Value Reporting Foundation. This move shall provide investors and corporates with a comprehensive corporate reporting framework across the full range of corporate value drivers and standards to drive global sustainability performance. "Information is the lifeblood of good decision making", says Robert K. Steel, Chair of the SASB Foundation Board of Directors, who continued commenting the merge in a statement "Capital markets are hungry for information linked to enterprise value creation, but they cannot easily digest what comes from a fragmented reporting landscape. This merger is an important step towards businesses and investors communicating with clarity and ease about the issues that matter most to financial performance".[10]

The purpose of the newly established Value Reporting Foundation is to maintain Integrated Reporting Framework, advocate integrated thinking, and sets sustainability disclosure standards for enterprise value creation. Since the IIRC Framework and the SASB Standards are complementary, the merger directly responds to calls from global investors and corporates to simplify the corporate reporting landscape, providing the market with a clear solution for communicating about the drivers of corporate value. "Integrated reporting describes all relevant value creation topics and the approach to integrating them in corporate thought and reporting. SASB provides the precise definitions of the data that should be reported for these topics in each industry. Organizations globally already use both to communicate effectively with investors about how sustainability issues are connected to long-term enterprise value, with these endeavours ultimately benefiting other key stakeholders. Under the Value Reporting Foundation, we will link the concepts between the <IR> Framework and SASB Standards even further", suggests Barry Melancon, Chair, IIRC Board.[11]

The merger follows the "Statement of Intent to Work Together Towards Comprehensive Corporate Reporting" released by the Carbon Disclosure Project, the Climate Disclosure Standards Board, the Global Reporting Initiative, the International Integrated Reporting Council and the Sustainability Accounting Standards Board where the key player of the non-financial and ESG disclosure landscape have come together to show a commitment to working towards a comprehensive corporate reporting system. Through this collaboration, these five parties intend to provide: a joint market guidance on how frameworks and standards can be applied in a complementary and additive way; a joint vision of how these elements could complement financial generally accepted accounting principles and serve as a natural starting point for progress towards a more coherent, comprehensive corporate reporting system; and a joint commitment to drive towards this goal, through an ongoing programme of deeper collaboration and a stated willingness to engage closely with other interested stakeholders. On 3rd November 2021 the IFRS Foundation Trustees announced the establishment of the International Sustainability Standards Board (ISSB) and the consolidation of the Value Reporting Foundation and the Climate Disclosure Standards Board (CDSB) into the IFRS Foundation. This move further consolidates the various initiatives addressing the "sustainability-related financial disclosures" aiming at providing investors and other capital providers of information that is relevant from the perspective of "enterprise value creation".

The Non-financial Reporting Directive

The 2014 Non-Financial Reporting Directive (NFRD)[12] has introduced in Europe for the first time an articulated set of requirements for the disclosures relating to matters that are generally referred to as *non-financial* in nature or relating to ESG aspects. It is also the first time that EU legislation requires that such non-financial information be provided as a self-contained set of disclosures which would together constitute a non-financial statement.

On 21 April 2021, the Commission adopted a proposal for a Corporate Sustainability Reporting Directive (CSRD), which would amend the existing reporting requirements of the NFRD. In the European Commission's proposal leading up to the current Directive, one of the key objectives of the new requirements was to: "*increase the transparency of certain companies, and to increase the relevance, consistency, and comparability of the non-financial information currently disclosed, by strengthening and clarifying the existing requirements*".[13] In pursuing this objective, the Directive would therefore help serving three main purposes: (i) incentivizing reporting companies to better manage the ESG factors and risks and opportunities associated with it; (ii) inform the public at large about the impacts arising from business operations on the environmental, social, and other aspects; and (iii) provide investors with more complete information on the long-term performance of companies.

The new requirements are introduced as amendments to the Accounting Directive[14] to which two articles were added—articles 19a and 29a—addressing, respectively, the non-financial statement and the consolidated non-financial statement that shall be prepared on an annual basis and included as part of an entity's management report, unless a separate report is prepared (as explained further in detail in paragraph "Non-Financial Reporting in the Context of Periodic Financial Information").[15]

It is immediately apparent for those familiar with the area of corporate reporting that a symmetry exists, in principle, between the non-financial information disclosed in a "non-financial statement" pursuant to the NFRD and the financial information traditionally included in the financial statements. Such symmetry, at least in theory, aims at responding to the growing interest by financial market participants, as well as other stakeholders, for complementing the traditional financial metrics—which have largely been accused of causing short-term pressures in financial markets—with qualitative and quantitative disclosures on an entity's value creation potential, its resilience, and its ability to act as a responsible *corporate citizen*.

In addition—and perhaps more importantly from a public policy perspective—the NFRD also aimed at achieving a broader objective, that is to enable financial market participants to channel funding and resources towards activities which are contributing to the shift towards a more sustainable financial system. From this perspective, the NFRD is only one piece of a larger jigsaw that currently sees multiple legislative acts, amongst which three main ones can be mentioned: the Taxonomy Regulation, the Disclosure Regulation (also referred to as Sustainable Finance Disclosure Regulation or SFDR), and the amendments to the Benchmarks regulation establishing two new types of benchmarks, i.e. the EU Climate Transition Benchmark and the EU Paris-aligned Benchmark.

Within the larger web of legislative acts relating to sustainable finance, the NFRD should, in principle, play a key role in providing basic data points for financial market participants to feed their investment decisions, but also to meet their own reporting obligations in accordance with the SFDR on the degree to which investment portfolios and products meet sustainable objectives. In practice, however, there is a significant lack of coordination between the NFRD requirements and the requirements in the SFDR and the Benchmarks regulation which limits the extent to which financial market participants and other actors can rely on NFRD disclosures to fulfil their own disclosure obligations.

The NFRD has also cross-sectorial implications given its applicability to a wide variety of financial market actors including corporates, asset managers, banks, and insurance undertakings. These factors explain why, in the past years, notwithstanding its limited scope, the NFRD and its application have been under increasing scrutiny by several parties: investors, corporates, NGOs as well as public authorities, including securities regulators.

There are, in particular, five main areas of the application of the NFRD that have attracted most attention by various stakeholders: (i) the scoping requirements of the Directive; (ii) the broader context surrounding the disclosures; (iii) the notion of materiality applied; (iv) the extent to which the level of disclosure has been regarded as adequate to meeting the information needs of various stakeholders; and (v) the areas for future improvement of the regime for non-financial reporting. As described in Table 4.1, in the next paragraphs of this section, we will address the first four aspects and leave the last one for discussion as part of Chapter 9.

The Scope of the NFRD

The NFRD addresses large undertakings which are public-interest entities exceeding on their balance sheet dates the criterion of the average number of 500 employees. Only for these entities, the co-legislators have intended to "*enhance the consistency and comparability of non-financial information disclosed throughout the Union*"[16] by requiring the preparation of a non-financial statement which shall address, as a minimum, information relating to environmental matters, social and employee-related matters, respect for human rights, anti-corruption, and bribery matters.

The Commission's proposal for a Directive on non-financial information explains this choice on the grounds of proportionality by indicating that "*further transparency should not translate into undue administrative burden. Smaller companies face more difficulties to collect and analyse information*". Such proportionality principle, which in the initial proposal aimed at targeting companies whose average number of employees during the financial year exceeds 500 and, on their balance sheet dates, exceed either a balance sheet total of EUR 20 million or a net turnover of EUR 40 million, has been further specified to ultimately introduce a narrower

Table 4.1 Mapping of the contents in this book relating to the NFRD and its future developments

Topic	*Chapter/Paragraph*
Scoping requirements	This chapter, section 'The Scope of the NFRD'
Broader context regarding non-financial information	This chapter, section 'Non-financial Reporting in the Context of Periodic Financial Information'
Notion of materiality	This chapter, section 'The Concept of Materiality in the Non-financial Statement and Its Application'
Adequacy of disclosures vis-à-vis stakeholders' needs	This chapter, section 'The Requirements to Disclose Non-financial Matters and Their Application'
Areas for future improvement for non-financial reporting	Chapter 9

scope by targeting not all companies which qualify as large undertakings under the Accounting Directive and exceeding the number of 500 employees, but only those undertakings that also qualify as public-interest entities.

Article 2 of the Accounting Directive defines a "public-interest entity" (PIE) as an entity listed in Annexes I and II of that Directive and falling in one of the following categories: (i) governed by the law of a Member State and whose transferable securities are admitted to trading on a regulated market of any Member State within the meaning of the MiFID II Directive[17] and excluding entities listed on a multilateral trading facility (these include, for example, entities listed on an SMEs growth market) or an organised trading facility; (ii) credit institutions as defined in Capital Requirements Directive IV[18] excluding certain—generally public sector-related—institutions, as referred to in Article 2 of that Directive; (iii) insurance undertakings as defined by Council Directive 91/674/EEC; and (iv) designated by Member States as public-interest entities, for example these include entities that are of significant public relevance because of the nature of their business, their size, or the number of their employees.

It shall be noted that not all of the listed entities in an EU-regulated market meeting certain dimensional criteria are required to report under the NFRD, but only those that are "governed by the law of a Member State", thus implying that a listed entity in an EU regulated market established outside the Union would not qualify as a PIE. It is also important to highlight that the scope of the NFRD includes entities that may qualify as a PIE based on specific national requirements. This latter element signals one of the multiple areas in which the NFRD allows for a certain degree of flexibility and adaptability to the national specificities, while giving rise to differences in the way the NFRD is applied amongst Member States thus making comparability and consistency across the Union more difficult to achieve as an objective. Another example of divergence in the application of the scoping requirements is the decision made by some Member States to extend the application of the requirements on non-financial information to PIEs with more than 250 employees.

As we will further discuss in Chapter 9, the choice to build the scope of the NFRD around dimensional criteria, adjusted by an element of public-interest relevance, could be regarded as a proxy of the impact that entities required to report under the Directive may have on (or bear from) the ESG factors. However, some concerns have been raised that the scope requirements may not necessarily reflect well the population of entities which may be most relevant from a sustainability perspective.

Non-financial Reporting in the Context of Periodic Financial Information

The Commission's proposal on the non-financial disclosures, which was then finalised as the NFRD, suggested that the new requirements should be incorporated into EU law by means of amendments to the Accounting Directive.

This choice was intended to allow for more flexibility in the way Member States could adapt the new requirements to their national specificities. In particular, the NFRD gives Member States the possibility to: (i) allow information relating to certain matters of particular commercial sensitivity to be omitted; (ii) exempt from the requirement to prepare the non-financial statement as part of the management report, an undertaking that prepares a separate report, provided that such report is published together with the management report or it is made publicly available on the undertaking's website within a reasonable period of time (no more than six months after the end of the reporting period) and it is referred to in the management report; and (iii) require that the information required by the NFRD be verified by an independent assurance services provider. As it could be expected, each of these options has given rise to differences in the transposition of the NFRD at the national level.[19]

The possibility of issuing a separate report on non-financial information is one of the requirements that poses both conceptual and practical questions. From a conceptual standpoint, in fact, the idea of providing disclosures in a single report, i.e. the management report, both from a financial and non-financial reporting perspective, seems appealing to improve the connectivity between financial and non-financial information. From a practical perspective, when considering how to maintain the length and complexity of the management report within a manageable range, the inclusion of an extensive set of non-financial disclosures does not seem to be always a feasible option. In a context in which users of corporate disclosures are expected to go through the disclosures in a booklet format, certainly the length of the documents may represent a concern. However, the increased use of digital reporting may relieve the pressure from considering this latter aspect as a critical one and it may *de facto* make it less relevant to know as part of which document—i.e. whether in the management report, the financial statements, or in a separate ESG report—certain data points are disclosed.

Whether the non-financial statement is part of the management report or provided as a separate report, an entity failing to provide in full or in part the required disclosures would contravene the requirement to prepare a management report of which, by default, the non-financial statement forms an integral part.

There is also another—more technical—reason for amending the Accounting Directive with the new non-financial reporting requirements: this Directive in fact establishes requirements which apply to a wide range of public and private limited liability companies. Therefore, the new requirements could be made applicable to a subset of this wide population as described in the previous paragraph, but also be better coordinated with the other types of periodic disclosures which this subset of entities typically prepare, i.e. the financial statements and the management report.

Both articles 19a and 29a of the Accounting Directive, in fact, acknowledge the connections existing with financial and non-financial information included in the financial statements and the management report. In particular, the NFRD requires that the non-financial statement (or the separate report) shall also include "where appropriate" references and additional explanations in relation to amounts reported in the annual financial statements. This link reflects one of the key challenges of non-financial information: trying and expanding the spectrum of information available to investors and other stakeholders not only to address ESG topics in isolation, but also to ensure that these new topics are contextualised within the broader set of disclosures of more traditional financial metrics and narrative explanations that are typically included in the financial statements, as well as the key performance indicators (KPIs), the analysis of risks and opportunities and disclosures on an entity's strategy that are typically provided in the management report.

In this respect, it shall be noted that Article 19 of the Accounting Directive—which pre-existed the introduction of the NFRD—already required to provide in the management report "*where appropriate*" non-financial KPIs including information on the environmental and employee matters. However, as a matter of fact, such information had not been widely incorporated into the reporting practices of most undertakings.[20]

It shall also be noted that, in line with the proportionality principle discussed in the previous paragraph, Article 19 also allows Member States to exempt small and medium-sized entities from the obligation to provide the above-mentioned non-financial KPIs. Given the wider scope of the requirements in Articles 19a and 29a of the Accounting Directive, it is considered that the provision of a non-financial statement is also sufficient to fulfil the requirement in Article 19 to disclose non-financial KPIs.

Another important aspect that is worth highlighting in relation to the interconnections between the new articles 19a and 29a and the requirements applicable to financial statements and management reports relates to the responsibility regime for drawing up the non-financial statement (whether as part of the management report or as a separate report). Article 33 of the Accounting Directive, in fact, requires that the members of the administrative, management, and supervisory bodies of an undertaking have collective responsibility for ensuring that the financial statements, the management report, including the non-financial statement and the separate report, shall be drawn up in accordance with the relevant reporting requirement (this requirement applies also to the corresponding consolidated versions of these documents).

If the choice of introducing the new non-financial reporting requirements into the Accounting Directive may have some theoretical benefits in terms of facilitating the connectivity with the contents of the financial statements and the management report, it shall be noted that the requirements for the non-financial statements in the Accounting Directive are not well connected with those in the Transparency Directive[21] and, in particular, with those concerning the periodic reporting obligations for listed entities.

In fact, Articles 19a and 29a of the Accounting Directive are not mentioned in Article 4 of the Transparency Directive that provides for the content and format of the annual financial report of issuers. This article, in fact, only cross-refers to Article 19 of the Accounting Directive, in relation to the management report *tout court*, thus raising the question on the reason behind this lack of coordination.

One of the consequences of this lack of coordination is that, in transposing the NFRD requirements at national level, not in all Member States there has been an automatic transfer of additional enforcement powers regarding the non-financial statements to the national authorities that were already competent at national level for the enforcement of the other reporting requirements included in the Transparency Directive.

Another consequence is that some confusion has arisen in practice as to whether the separate report that would replace the non-financial statement included within the management report falls within the broader reporting requirements of Article 4 of the Transparency Directive. In this respect, it would seem reasonable to assume that, even when a separate report is prepared based on an exemption provided at national level pursuant to paragraph 4 of Articles 19a or 29a of the Accounting Directive, such separate report shall still be reconnected to the primary obligation to include a non-financial statement in the management report. On this basis, if the separate report were not to be fully compliant with the requirements for the contents of the non-financial statements, any omissions or misstatements should be indirectly regarded as lack of compliance with the content requirements for the management report which is one of the mandatory documents to be included in the annual financial report based on the Transparency Directive. However, such an indirect link is not explicit neither in the Transparency Directive nor in the Accounting Directive.

The Concept of Materiality in the Non-financial Statement and Its Application

One of the key aspects in the application of the NFRD is also its key innovation: the concept of "double materiality". This concept is introduced in the Accounting Directive in the first paragraph of Articles 19a and 29a, and it can be derived from the reference in these articles to the fact that entities shall disclose (emphasis added) "*information to the extent necessary for an understanding of the undertaking's development, performance, position and impact of its activity, relating to, as a minimum*" the non-financial matters mentioned in the NFRD.

Based on these requirements, not only it is necessary to disclose information on how the performance and position of the entity are affected by the non-financial matters, but it is also necessary to disclose how the actions of an entity may have consequences vis-à-vis environmental, social, and governance factors and any other matters that may broadly qualify as non-financial information. This *dual lens* through which materiality of the information to be reported is assessed, is the reason for which this notion is usually referred to as "double materiality".

The European Commission's 2017 non-binding guidelines on non-financial information have been particularly effective in describing this notion by explaining that the NFRD: "introduced a new element to be taken into account when assessing the materiality of non-financial information. In effect, the Non-Financial Reporting Directive has a double materiality perspective". The first *direction* of materiality is "outside-in", i.e. the so-called *financial materiality* and relating to actions and facts that affect the value of a company. The second direction is "inside-out" and refers to the *environmental and social materiality*.

The neat distinction between the two directions of the double materiality assessment does not exclude the fact that, in practice, significant connections exist between these two aspects. For example, the impacts that an entity has on its external environment may give rise to a sort of *rebound effect*, whereby that entity will bear the consequences arising from, for example, the unfair treatment of its employees, the pollution of the surrounding environment, or the disrespect for human rights. Such feedback effect may arise from reputational or legal risks as well as from climate-related physical risks which a certain company may not have contributed to mitigate or that it has even contributed to exacerbate.

The rebound effect is sometimes seen by some as the maximum possible extension of the financial materiality boundary and the only part of the "inside-out" perspective that is of relevance for investors. However, the growing interest in impact investing seems to signal that for investors the relevance of the "inside-out" perspective does not seem to be limited only to the rebound effect.

This double materiality principle co-exists with the definition of "material" that was already included in the Accounting Directive before the introduction of the non-financial statement and which generically refers to material information as: "*the status of information where its omission or misstatement could reasonably be expected to influence decisions that users make on the basis of the financial statements of the undertaking. The materiality of individual items shall be assessed in the context of other similar items*".

However, this notion is not meant to be applied to the information included in the non-financial statement, but only to financial statements. This implies that one important element of this definition does not necessarily apply to non-financial disclosure, i.e. the reference to the fact that any omissions or misstatements of information, in order to qualify as referring to material information, shall be *reasonably expected to influence the decisions of users of that information*. In other words, users may take different decisions (for example, to invest or not to invest, to lend resources or not to do so, etc.) depending on the information made available to them.

We note that the lack of this important reference in the notion of materiality for non-financial information may be one of the key elements that make the double materiality notion appealing in theory, but also rather difficult to apply in practice without the support of clear reporting standards and even potentially prone to manipulation in some cases. It could be argued, in fact, that especially the "inside-out" aspect of materiality may raise one question: "*Where do we stop with the identification of the impacts that an entity may have on the surrounding environment?*" or "*Over which horizon is it relevant to inspect the impacts of companies on the ESG factors?*".

In practice, many entities preparing non-financial statements develop so-called "materiality matrices" based on multi-dimensional assessments and stakeholder questionnaires or leveraging on more sophisticated IT tools to identify material topics. Typically, these assessments evolve with time together with the awareness of companies and the public at large thanks to both cultural and regulatory change. Some have referred to this evolving nature of materiality as "dynamic materiality"[22] to describe the changing nature of this concept and the processes which can be put in place to capture on a timely basis the topics that qualify as material.

Whichever approach is adopted in practice to determine what qualifies as material information, it shall be noted that the definition of what is material in the non-financial area is critical to determine what disclosures entities shall provide to fulfil the broad information categories required by the NFRD. From this perspective, it could be argued that for practitioners the notion of materiality in non-financial reporting may have an even more central role compared to that played by this concept in financial reporting, where detailed reporting requirements establish the contents of the financial disclosures.

This results in a fundamental difference in the degree of comparability of the information provided in financial statements compared to non-financial statements. In the absence of mandatory standards, in fact, a general notion of materiality alone has proven to be ineffective in providing comparable, relevant, and reliable disclosures on ESG factors. Nevertheless, the double materiality notion constitutes a basic building block of the EU disclosure regime for non-financial information and one which will certainly need to be further specified in the forthcoming revision of the NFRD, as discussed further in Chapter 9.

In this context, it is noteworthy that major international standard-setters have already developed a quite substantial amount of experience in the implementation of different materiality notions, most notably a major distinction can be made between the SASB's materiality concept and that of GRI. In its 2017 Conceptual Framework (currently under review),[23] SASB indicates that its standards "*address the sustainability topics that are reasonably likely to have material impacts on the financial condition or operating performance of companies in an industry*" and that "*In identifying sustainability topics that are reasonably likely to have material impacts, the SASB applies the definition of 'materiality' established under the U.S. securities laws. According to the U.S. Supreme Court, information is material if there is 'a substantial likelihood that the disclosure of the omitted fact would have been viewed by the reasonable investor as having significantly altered the "total mix" of information made available'*". In their 2020 Exposure Draft with proposals to revise the existing Conceptual Framework, SASB amended the notion of materiality to more explicitly define financial materiality and more clearly explain that this notion also embeds the one of the enterprise value creation, as follows: "*For the purpose of SASB's standard-setting process, information is financially material if omitting, misstating, or obscuring it could reasonably be expected to influence investment or lending decisions that users make on the basis of their assessments of short-, medium-, and long-term financial performance and enterprise value*". On the other hand, GRI as part of its reporting principles indicates that a sustainability report prepared on the basis of its standards shall: "*cover topics that: reflect the reporting organization's significant economic, environmental, and social* ***impacts****; or substantively influence the assessments and decisions of* ***stakeholders***".[24] The EU notion of double materiality incorporates elements of both those aspects without explicitly referring to any of those sets of standards in particular. We will discuss in Chapter 9 the challenges associated with ensuring compatibility and interoperability across the different notions of materiality adopted by different reporting regimes.

The Requirements to Disclose Non-financial Matters and Their Application

The NFRD focusses on six key non-financial matters:

- the environment;

- social aspects;
- employee-related aspects;
- respect for human rights;
- anti-corruption; and
- bribery-related issues.

This list is intended to identify the areas which, as a minimum, need to be addressed, but other matters depending on their materiality may also need to be disclosed by entities. However, the NFRD does not provide specific examples of any such additional topics.

In cases where an entity indicates that it does not pursue a policy in relation to one or more of these matters, the non-financial statement shall provide a clear and reasoned explanation for not doing so. This disclosure—which some regard as a form of *comply-or-explain* approach—is too often disregarded or only formally fulfilled with information that is not entity-specific and *boilerplate*. It shall also be noted that in requiring this disclosure, the NFRD does not distinguish between a situation where an entity does not pursue a policy relating to a certain non-financial matter based on materiality considerations or due to other reasons.

For each of these aspects, the NFRD requires information around five main categories:

- the entity's business model;
- the policies pursued by the entity and the associated due diligence processes;
- the outcome of those policies;
- the principal risks related to those matters and linked to the entity's operations; and
- the non-financial key performance indicators relevant to the concerned business.

The observation of the reporting practice since the application date of the new requirements in 2017 shows that while entities tend to report information on each of the above-mentioned five categories, the degree to which these disclosures provide useful insights into the entity's targets and the progress made to meeting them varies significantly and the risk is that some entities interpret the non-financial disclosures as a mere compliance exercise.

For example,[25] while the disclosure of the business model is generally the one where most reporting entities tend to provide an extensive description, in substance the links with the non-financial matters are missing. In fact, when describing the business model is not always clear how the strategic orientation of an entity reflects or is impacted by the non-financial matters mentioned in the NFRD.

Similarly, the risk reporting required by the Accounting Directive as part of the management report is often disconnected from the reporting of risks arising from the non-financial matters. In some cases, the risk reporting in accordance with the NFRD merely refers to generic risks associated to ESG factors, without explaining how such risks are specifically impacting an entity's business model and what mitigating actions, if any, are put in place by the management. In other cases, the disclosures only refer to the risks that entities generate in relation to the non-financial matters with an impact on the public at large (the inside-out perspective), but they omit to disclose how the non-financial risks may impact the entity itself (the outside-in perspective).

Another example relates to the disclosures of the non-financial policies which are generally provided by most entities, even though the level of specificity of such disclosures is more extensive for environmental issues, including climate change, and far less widespread on matters relating, for instance, to human rights and anti-corruption. In any case, even in the areas where policies are more extensively discussed, the description of the outcomes of such policies and the risks arising from non-financial matters is non-existent or often very generic.

Another important topic for attention is the level of disclosures relating to due diligence processes associated with the non-financial matters. Such disclosures should provide insights into the internal corporate governance arrangements (e.g. at the board and senior management level) that are in place to ensure the effective management by the reporting entities of the non-financial matters and the related policies which they declare to commit to. Unfortunately, the current practice does not seem to show a satisfactory level of transparency in this area, especially when it comes to assessing the non-financial matters along an entity's supply chain. This is particularly, although not solely, important for social and employee matters and for the assessment of policies relating to the respect for human rights.

Obviously the examples provided in the previous paragraphs give a snapshot on the quality of the reporting practices which then need to be contextualised depending on the different sectors that are reviewed. Some granular assessments provided in various public reports published in the last years shed light on these aspects in more detail, but at least three general conclusions can be drawn in relation to the status of non-financial reporting at the time of writing this chapter.

Firstly, even for non-financial matters, such as climate change, which attract a significant level of public scrutiny also by European legislators, the quality of corporate disclosures by European entities can be significantly improved. This is consistent with a broader trend at international level where, for example, the Task Force for Climate-Related Financial Disclosures (TCFD), in its 2020 status report notes that: "companies' disclosure of the potential financial impact of climate change on their businesses and strategies remains low [...]".[26] In its recent report on "The state of EU Environmental Disclosure in 2020"[27] in which the 2020 environmental disclosures of 50 of Europe's largest listed companies were reviewed, the Climate Disclosures Standards Board (CDSB) notes that the "*2020 analysis still shows TCFD adoption to be inconsistent and incomplete. Whilst 68% of company disclosures now make some reference to TCFD, the vast majority have still only partially adopted the recommended disclosures. For example, 96% of companies still do not define short, medium and long-term time horizons, and just 18% are providing clear disclosure on their resilience to different climate scenarios*".

Secondly, there is certainly an issue associated with the stage of the learning curve which different reporting entities are at. In fact, some entities might be more advanced than others, also thanks to the higher number of resources committed to the green transition and this factor, coupled with the absence of sufficiently detailed requirements and standards on the information to be disclosed and the methodological guidance to follow, may explain the limited quality of disclosures. At the same time, we shall note that investors' demand for non-financial disclosures is increasing due to regulatory pressure, as well as to higher public awareness in relation to ESG-related topics. In this context, providers of ESG data, scoring, ratings, and other ESG assessment tools have seen the demand for their products increasing dramatically, while the quality of the source data from reporting companies which feeds their databases has not necessarily improved at the same pace. The combination of these factors creates a "perfect storm" for corporate disclosures to become the channel of transmission from issuers to investors of information that is potentially misleading with the risk of having sub-optimal capital allocation decisions, detracting from the credibility of the sustainable finance initiative and creating the conditions for new undue short-term pressures in the financial markets.

Thirdly and lastly, different actors in the corporate reporting ecosystem currently require a more stringent and homogeneous regime across Europe. Companies need to report on the basis of high-quality reporting standards; auditors need to provide assurance on the contents of the non-financial disclosures provided; and the mandate of supervisors with regards to non-financial reporting needs to be strengthened and harmonised across Europe so to ensure performance efficient and effective enforcement of non-financial disclosures. In other words, it is unlikely that significant improvement in the quality of non-financial information will be achieved without intervening with clear requirements to reinforce the key role that each of these actors must play in the corporate reporting chain.

One of these interventions took place recently, as the Shareholder Rights Directive was revised in an attempt to reinforce long-term investment and engagement. We further explore this within the next section.

Corporate Disclosure and Long-Term Shareholder Engagement: The Review of the Shareholder Rights Directive and the New Provisions in the Area of Directors' Remuneration

The June 2019 entry into application of the revised Shareholder Rights Directive (SRDII) has introduced a number of disclosure duties aimed at facilitating long-termism by means of better-informed shareholder monitoring and more active engagement vis-a-vis listed companies. One of the key assumptions at the basis of the SRDII is that shareholders' monitoring role is insufficient and their engagement should be reinforced to mitigate companies' excessive focus on short-term returns.[28]

As a background, it should be noted that (institutional) investors with long-term liability structure and investment horizon, such as pension funds, nowadays collectively hold a substantial fraction of the equity market.[29] As such, their engagement with investee companies may have substantial effects on how such companies are managed, including their strategy and how they approach sustainability issues. While the traditional corporate finance literature argues that (dispersed) shareholder engagement is hampered by structural free-riding issues (so-called rational apathy), in recent years there has been an increasing expectation that certain institutional investors can steer firms towards long-term value creation and social welfare maximization rather than simply pursuing short-term market value.

Although this issue remains highly controversial, the literature has found some evidence of the beneficial role of engagement in terms of increasing shareholder value over a long-term time frame.[30] At the same time, there is evidence that shareholder pressure is increasingly focusing not only on governance but also on environmental and social issues, i.e. all the three ESG components have become material issues for shareholder engagement.[31] As such, efforts made by some institutional investors in the direction of engaging with the companies they invest in so to voice their concerns have been welcomed and facilitated by the regulatory framework, both at the national[32] and EU level.

To this goal, the SRDII legislates in a number of areas affecting different players in the financial market. For example, Art. 3 (c) provides that intermediaries should facilitate the exercise of the rights by the shareholder, including the right to participate and vote in general meetings. Article 3 (g) mandates asset owners and asset managers to develop and publicly disclose an engagement policy that explains how they monitor investee company and exercise voting rights on relevant matters, including strategy, financial and non-financial performance and risks, capital structure, social and environmental impact and corporate governance. Institutional investors and asset managers should report annually on how this policy has been implemented. In addition, Article 3 (j) sets out disclosure requirements for proxy advisors, also with reference to their adherence to a code of conduct. Finally, Article 9 (a) (b) provides that companies should draw a clear and understandable remuneration report and grants shareholders the right to vote on the remuneration policy.

In the following, this paragraph will look into the SRDII new provisions that may have the highest impact on corporate disclosures related to ESG factors, namely those in the area of directors' remuneration.

Directors' Remuneration: The Rationale for Policy Measures

Directors' remuneration is certainly a core area of shareholder monitoring due to the important role that remuneration has in shaping managerial incentives and—in turn—companies' long-term strategy. By the same token, directors' remuneration is also a key area of corporate disclosure as it provides to the outside certain pieces of information that are extremely relevant in showing the managerial commitment towards long-term value creation and ESG goals.

The rationale underpinning policy measures in this area is that increased companies' disclosure together with a strengthened ability of shareholders to monitor and express their voice may provide for better accountability of directors and—in turn—more focus on the long-term interests of the company and sustainability. As mentioned, the academic literature traditionally argues that shareholders of companies with dispersed ownership bear high information asymmetries that make monitoring of their management challenging. Due to limited monitoring, executives may be prone to focus on personal interests rather than those of the company and its shareholders.[33] Remuneration can therefore be an important tool for the shareholders to steer managerial efforts towards sustainable value creation. However, depending on how remuneration packages are designed, they could also skew managerial interests towards short-term value, given that remuneration packages are often linked to short-term metrics such as yearly returns or equity performance.[34]

In this context, the SRD II explicitly calls for a remuneration policy that is not linked to short-term objectives but rather focusses on the company's long-term interests and sustainability.[35] In addition, the SRD II requires that directors' performance is assessed not only based on financials but also by means of non-financial performance criteria that should include—where appropriate—ESG factors.[36] This is in line with other policy initiatives at the international level, including the 2015 G20/OECD Corporate Governance Principles.[37] A part of the academic literature confirms that if remuneration is linked to well-developed metrics of long-term sustainability, it can indeed be crucial in better aligning directors' incentives with those of the society at large.[38]

New SRD II Provisions in the Area of Directors' Remuneration

The SRD II addresses the area of directors' remuneration in listed companies[39] in three ways:

(i) Firstly, it mandates the publication of a remuneration policy that shall contribute to the company's *business strategy and long-term interests and sustainability* (Art. 9 (a) (1) (6));
(ii) Secondly, it provides for the content of the remuneration report that should comply with the remuneration policy and its goal to contribute to the long-term performance of the company (Art. 9 (b) (1));
(iii) Thirdly, it requires a shareholder vote both on the remuneration policy and on the remuneration report (Art. 9 (a) (1) (2) (3) (5) and Art. 9 (b) (4)).

(i) As mentioned, Article 9 (a) (1) (6) provides that companies publish a remuneration policy as regards directors and that such policy explains how it contributes to *the company's business strategy and long-term interests and sustainability*. This document should be *clear and understandable* and *describe the different components of fixed and variable remuneration, including all bonuses and other benefits in whatever form* that can be awarded to directors.

In addition, Art. 9 (a) (6) establishes that companies should set clear, comprehensive, and varied criteria for the award of variable remuneration, distinguishing between *financial and non-financial performance criteria*. These latter should explicitly include, when appropriate, *criteria relating to corporate social responsibility*. Further details should be disclosed on the application of performance criteria, the functioning of deferral periods or claw-back clauses. In case of share-based remuneration, companies are required to disclose specifics such as the vesting period and any retention of shares after such period.

Finally, disclosure is mandated on the decision-making process leading to the design and implementation of the remuneration policy as well as its governance, both with reference to the role of the remuneration committee and how AGM views have been taken into account.

(ii) Coming to the remuneration report, its content is provided by Art. 9 (b) (1) clarifying the minimum information items needed to provide a comprehensive overview of the remuneration assigned to companies' directors, either executive or non-executive.

Remuneration includes both that awarded and due during the most recent financial year and all benefits in whatever form. Further details are mandated with reference to the components of remuneration and it is explicitly provided that a proper explanation should be given on how the remuneration *effectively contributes to the long-term performance of the* company, in line with the adopted remuneration policy, and how performance criteria were applied. Any deviation or derogation to the policy should be carefully explained.

Finally, some information is required with reference to stock options and the main conditions for their exercise as well as the changes of directors' remuneration in comparison to the year before and in connection with the company's performance and the evolution of the average remuneration of companies' employees.

(iii) Ex-ante disclosure of the remuneration policy and ex-post disclosures included in the remuneration report are supported by the introduction of ad-hoc AGM votes.

Art. 9 (a) (2) (3) (5) specifies that the vote on the remuneration policy should take place at least every four years and is by default binding.[40] While in the absence of an approved policy the company may continue paying remuneration to its directors, a revised policy must be submitted to the following AGM. It should be noted that temporary and exceptional circumstances may allow companies to derogate from the remuneration policy in case that is *necessary to serve the long-term interests and sustainability of the company* as a whole or assure its viability. With reference to the remuneration report, Art. 9 (b) (4) provides for an advisory AGM vote[41] and mandates the company to explain in the report how the previous year's AGM vote has been taken into account.

The regulatory framework set out in the SRD II may be complemented soon by guidelines published by the Commission. Such guidelines, that are going to be issued based on the mandate of Art. 9 (b) (6) of the SRDII, will specify the standardized presentation of the remuneration report. The draft guidelines published by the Commission in March 2019 address long-term sustainability by requiring that the remuneration report explains how the assigned remuneration complies with the relevant remuneration policy and in such way contributes to the *long-term interests and sustainability of the company*.[42] The draft Commission guidelines further elaborate on how the report could shed light on the consistency between the criteria the company has used in its own remuneration policy, the performance achieved over the reported financial year and the actual awards resulting from each criterion in a way that allows to distinguish between one-year and multi-year incentives.

Overall, the SRD II package on directors' remuneration provides a thorough set of measures that may improve corporates' focus on the long-term via enhanced transparency and strengthened shareholder monitoring. However, it will be crucial to observe whether the new set of rules will be effective in making market practice evolve and in particular whether ESG factors will effectively be commonly used to assign part of the variable remuneration as well as if the relevant disclosures will be effective in allowing shareholders to easily identify potential best practices that are going to be developed.

Here, it should be noted that while issuers may need time to develop ESG remuneration practices, the time needed to complete the transition will be most likely shortened if institutional investors exercise pressure via enhanced engagement on remuneration practices (see also Chapter 7). As argued by ESMA,[43] increasingly standardized and tabular disclosure might provide incentives for investors to support the development of market practice towards the use of ESG parameters as KPIs to which remuneration can be effectively linked. In this context, the content of the guidelines to standardize the presentation of the remuneration to be issued by the Commission under Art. 9 (b) (6) will be key, as well as their application at MS level. In addition, the recently proposed Corporate Sustainability Due Diligence Directive (COM/2022/71) might play a role in this area, as further explained in Chapter 9.[44]

Final Remarks Concerning ESG Corporate Disclosures

Within this chapter we have attempted to explore the corporate disclosure landscape as non-financial information becomes paramount. We started by summarizing the recent evolutions in the unfolding scenario of non-financial reporting and ESG disclosure. Then, we reflected on two important recent developments that have impacted on the European corporate disclosure landscape, namely, the Non-Financial Reporting Directive and the revised Shareholder Rights Directive.

Looking at the main pillars of non-financial and ESG disclosures under the existing legislation, we have been trying to adopt a cross-sectoral approach and bridge the gaps between different legislative sources. In this context, it is relevant to observe that, while the standardized presentation of ESG disclosures is clearly of utmost importance to ensure a smooth and quick transition towards improved market practice, the outputs of such disclosure, i.e. the relevant reports and documents that include that, are often inconsistent in terms of structure, availability, and assurance.

Some of the most relevant ESG disclosures are included in the corporate governance report, the non-financial statement, and the remuneration report. However, these documents are not always included in the management report and, in turn, in the financial statement. On the contrary, they are often disclosed separately in those MS that opted-out from setting a single disclosure document for all the information that is not included in the financial statement. This may have an impact on the effective accessibility of ESG-related information across the spectrum of corporate disclosures, with consequences in terms of clarity and consistency that should not be overlooked.

Taken in isolation, the ESG-related information currently provided in the non-financial statement, in the corporate governance statement, and in the remuneration report provides helpful disclosure to users of these statements. However, the split of ESG-related information across different documents risks creating inconsistencies and overlaps in the disclosure. In this context, harmonizing the location of the above three documents would help users of ESG-related information to access this in the same place. Should that location be the management report, that would also reinforce the level of assurance, both in terms of public and private enforcement. On the flipside of the coin, placing such information outside the management report (possibly referenced via a hyperlink) might provide better incentives for companies to provide more ESG-related information than that which fits in the management report, thereby facilitating the development of best practices in this area.

Notes

1. Grice, A., Reeves, M., and Fuller, J., "Getting Uncomfortable on Purpose," Boston Consulting Group and BCG Henderson Institute, January 23, 2019, https://www.bcg.com/ru-ru/publications/2019/getting-uncomfortable-on-purpose.aspx.
2. See the 2018 Larry Fink's letter to CEOs, January, https://www.blackrock.com/corporate/investor-relations/2018-larry-fink-ceo-letter.

3. Some of these reflections were also offered on Busco, C., Eccles, R. G., Consolandi, C., and Sofra, E., "A Preliminary Analysis of SASB Reporting: Disclosure Topics, Financial Relevance, and the Financial Intensity of ESG Materiality", 2020, *Journal of Applied Corporate Finance*, Volume 32/2, pp. 117–125.
4. Eccles, R. G., and Klimenko, S., "The Investor Revolution. Shareholders Are Getting Serious About Sustainability," *Harvard Business Review*, April 13, 2019, https://hbr.org/2019/05/the-investor-revolution.
5. See https://globalreporting.org/.
6. See https://integratedreporting.org/.
7. See https://www.sasb.org/.
8. In an interview, Tim Mohin, CEO of GRI, addressed the relationship between GRI and SASB [Allison-Hope, D., "The Sustainability Reporting Landscape: Q&A with GRI and SASB," *BSR*, November 30, 2017, https://www.bsr.org/en/our-insights/blog-view/ sustainability-reporting-landscape-interview-gri-sasb]. Janine Guillot, chief executive of SASB, says the standard setters are not in competition. "SASB is uniquely designed for investors," she says. "On top of that the world needs a broader set of standards on how companies affect the world, and that's where the GRI comes in" [Mooney, A., and Nauman, B., "Larry Fink Rules on the Best Global Standards for Climate Risk Reporting," *Financial Times*, January 20, 2020, https://www.ft.com/content/fc51227b-9d64-4e5a-b1e2-f6c07f4caa58].
9. See the 2020 Larry Fink's letter to CEOs, January, https://www.blackrock.com/corporate/investor-relations/larry-fink-ceo-letter.
10. See "IIRC and SASB announce intent to merge in major step towards simplifying the corporate reporting system", on the IIRC web site (retrieved 28 November 2020).
11. See "IIRC and SASB announce intent to merge in major step towards simplifying the corporate reporting system", on the IIRC web site (retrieved 28 November 2020).
12. Directive 2014/95/EU of the European Parliament and of the Council of 22 October 2014 amending Directive 2013/34/EU as regards disclosure of non-financial and diversity information by certain large undertakings and groups. In 2020, the European Commission published a comprehensive study prepared by CEPS on the NFRD and its application which is available here. Different other studies have assessed the transposition of the NFRD across EU countries, *ex multis* see: Testarmata, S., Ciaburri, M., Fortuna, F., and Sergiacomi, S., "Harmonization of Non-financial Reporting Regulation in Europe: A Study of the Transposition of the Directive 2014/95/EU". In: Brunelli, S., and Di Carlo E. (eds) *Accountability, Ethics and Sustainability of Organizations*. Accounting, Finance, Sustainability, Governance & Fraud: Theory and Application.

Springer, 2020. Aureli, S., Salvatori, F., and Magnaghi, E., "A Country-Comparative Analysis of the Transposition of the EU Non-Financial Directive: An Institutional Approach", *Accounting, Economics, and Law: A Convivium*, vol. 10, no. 2, 2020. Noti, K., Mucciarelli, F. M., Angelici, C., dalla Pozza, V., and Pillinini, M., "Corporate Social Responsibility (CSR) and Its Implementation into EU Company Law", Policy Department for Citizens' Rights and Constitutional Affairs Directorate-General for Internal Policies and commissioned by the European Parliament, 2020. CSR Europe and GRI, in collaboration with Accountancy Europe, Member State Implementation of Directive 2014/95/EU, 2018.

13. Proposal for a Directive of the European Parliament and of the Council amending Council Directives 78/660/EEC and 83/349/EEC as regards disclosure of non-financial and diversity information by certain large companies and groups /* COM/2013/0207 final—2013/0110 (COD).
14. Directive 2013/34/EU of the European Parliament and of the Council of 26 June 2013 on the annual financial statements, consolidated financial statements and related reports of certain types of undertakings, amending Directive 2006/43/EC of the European Parliament and of the Council and repealing Council Directives 78/660/EEC and 83/349/EEC.
15. In this chapter, unless otherwise stated, when the term "non-financial statement" is used it is intended to refer to also to the separate report which, according to the NFRD, can replace the non-financial statement.
16. Cfr. Recital 6 of the NFRD.
17. Cfr. point (21) of Article 4 (1) of Directive 2014/65/EU.
18. Cfr. point (1) of Article 3 of Directive 2013/36/EU.
19. Insert reference to Accountancy Europe report.
20. In this respect, the Commission' proposal indicates that: "consultations have shown that only a limited number of EU large companies regularly disclose non-financial information, and the quality of the information disclosed varies largely".
21. Directive 2004/109/EC of the European Parliament and of the Council of 15 December 2004 on the harmonization of transparency requirements in relation to information about issuers whose securities are admitted to trading on a regulated market and amending Directive 2001/34/EC.
22. Kuh, T., Shepley, A., Bala, G., and Flowers, M., "Dynamic Materiality: Measuring What Matters".
23. Available at: https://www.sasb.org/wp-content/uploads/2020/02/SASB_Conceptual-Framework_WATERMARK.pdf.
24. GRI, GRI 101: Foundation 2016, paragraph 1.3. GRI standards can be downloaded here. The GRI Universal Standards are under review and the revised version is expected in 2021.
25. Alliance for corporate transparency, report Review of 1000 companies, https://www.allianceforcorporatetransparency.org/assets/2019_Rese

arch_Report%20_Alliance_for_Corporate_Transparency-7d9802a0c18c9f13017d686481bd2d6c6886fea6d9e9c7a5c3cfafea8a48b1c7.pdf.
26. TCFD status report, https://www.fsb.org/wp-content/uploads/P291020-1.pdf, p. 3.
27. The state of EU environmental disclosure in 2020, https://www.cdsb.net/sites/default/files/cdsb_eu_environmental_disclosure_in_2020.pdf.
28. In line with this approach, the 2012 Kay Review highlighted that the less the shareholders are monitoring the company, the more the firm tends to undertake actions such as frequent internal reorganization, inorganic corporate strategies on mergers and acquisitions and financial re-engineering that might reduce firms' and shareholders' value.
29. For recent aggregate data on institutional investors' ownership in listed companies across the world, please see Fig. 2.34 in OECD (2021), *The future of corporate governance in capital markets following the Covid-19 crisis*.
30. For example, Gompers et al. (2003) illustrated that, over a 10-y horizon, firms that are the target of robust investor engagement that improves their corporate governance outperform other firms by 8.5% per year. More recent research in this area includes Kölbel et al. (2020), Elst (2018), Dimson et al. (2015), Barko et al. (2017), Bauer et al. (2013), Hoepner et al. (2018), Motta and Uchida (2018).
31. FRC (2020), ISS (2014), Dimson et al. (2015).
32. One key example in this area are the several regulatory measures put in place in the UK following the Kay Review.
33. Bebchuk and Fried (2003).
34. Cf. for example Salazar and Mohamed (2016).
35. Cf. for example Recital 28 of the SRD II: "Directors contribute to the long-term success of the company. (…) Since remuneration is one of the key instruments for companies to align their interests and those of their directors and in view of the crucial role of directors in companies, it is important that the remuneration policy of companies is determined in an appropriate manner by competent bodies within the company and that shareholders have the possibility to express their views regarding the remuneration policy of the company".
36. Cf. Recital 29 of the SRD II: "(…) The remuneration policy should contribute to the business strategy, long-term interests and sustainability of the company and should not be linked entirely or mainly to short-term objectives. Directors' performance should be assessed using both financial and non-financial performance criteria, including, where appropriate, environmental, social and governance factors. The remuneration policy should describe the different components of directors' pay and the range of their relative proportions. (…)".

37. The G20/OECD Principles of Corporate Governance help policy makers evaluate and improve the legal, regulatory, and institutional framework for corporate governance. They also provide guidance for stock exchanges, investors, corporations, and others that have a role in the process of developing good corporate governance. First issued in 1999, the Principles have become the international benchmark in corporate governance. They have been adopted as one of the Financial Stability Board's Key Standards for Sound Financial Systems and endorsed by the G20.
38. Cf. for example Salazar and Mohamed (2016). It should be noted however that there are several papers putting into question the ability of remuneration disclosure (Barontini et al., 2013) and AGM voting (Fisch et al., 2018) to favour per se the achievement of long-term value maximization.
39. It should be noted that—differently from the NFRD—the scope of application of the SRD II includes all listed companies.
40. It should be noted that—differently from the original Commission's proposal—Member States may opt-out from this provision and make the vote advisory.
41. A specific carve out is designed for small and medium enterprises that are provided by Art. 9 (b) (4), second paragraph, with an enhanced level of flexibility.
42. Guidelines on the standardised presentation of the remuneration report under Directive 2007/36/EC, as amended by Directive (EU) 2017/828 as regards the encouragement of long-term shareholder engagement, March 2019.
43. ESMA Report on undue short-term pressure on corporations from the financial sectors, December 2019.
44. EC (2022).

References

Barko, T., Cremers, K. J. M., & Renneboog, L. (2017). *Shareholder engagement on environmental, social, and governance performance* (CentER Discussion Paper Series No. 2017-040, European Corporate Governance Institute (ECGI)—Finance Working Paper No. 509/2017, TILEC Discussion Paper No. DP 2017-021). https://ssrn.com/abstract=2977219

Barontini, R., Bozzi, S., Ferrarini, G., & Ungureanu, M. C. (2013). Directors' remuneration before and after the crisis: Measuring the impact of reforms in Europe. In M. Belcredi & G. Ferrarini (Eds.), *Boards and shareholders in European listed companies: Facts, context and post-crisis reforms*. Cambridge University Press.

Bauer, R., Clark, G. L., & Viehs, M. (2013). *The geography of shareholder engagement: Evidence from a large British institutional investor.* https://doi.org/10.2139/ssrn.2261649

Bebchuk, L. A., & Fried, J. M. (2003). Executive compensation as an agency problem. *Journal of Economic Perspectives, 17*(3), 71–92.

Dimson, E., Karakas, O., & Li, X. (2015). Active ownership. *Review of Financial Studies, 28*, 3225–3268 (Fox School of Business Research Paper No. 16-009).

EC. (2022). Proposal for a Directive of the European Parliament and of the Council on Corporate Sustainability Due Diligence and amending Directive (EU) 2019/1937, available at https://ec.europa.eu/info/sites/default/files/1_1_183885_prop_dir_susta_en.pdf

Elst, C. (2018). Shareholder sustainability activism in the Netherlands, European Corporate Governance Institute (ECGI) (Law Working Paper No. 396).

Financial Reporting Council (FRC). (2020). The UK Stewardship Code.

Fisch, J. E., Palia, D., & Solomon, S. D. (2018). Is say on pay all about pay? The Impact of firm performance. *Harvard Business Law Review.*

Gompers, P., Ishii, J., & Metrick, A. (2003). Corporate governance and equity prices. *Quarterly Journal of Economics, 118*, 107–155.

Hoepner, A. G. F., Oikonomou, I., Sautner, Z., Starks, L. T., & Zhou, X. Y. (2018). ESG shareholder engagement and downside risk, AFA 2018 paper, European Corporate Governance Institute (Finance Working Paper No. 671/2020).

ISS for the Investor Responsibility Research Center Institute. (2014). *Defining engagement: An update on the evolving relationship between shareholders, directors and executives.*

Kölbel, J., Heeb, F., Paetzold, F., & Busch, T. (2020). *Can sustainable investing save the world? Reviewing the Mechanisms of Investor Impact in Organization & Environment*. https://ssrn.com/abstract=3289544

Motta, E. M., & Uchida, K. (2018). Institutional investors, corporate social responsibility, and stock price performance. *Journal of the Japanese and International Economies*, Elsevier, 47(C), 91–102.

Salazar, A., & Mohamed, M. (2016). *The duty of corporate directors to tie executive compensation to the long-term sustainability of the firm* (Osgoode Hall Law School, Legal Studies Research Paper No. 20, Volume 12, Issue 4).

CHAPTER 5

Third Parties Information

Mario La Torre, EscrigOlmedo Elena, Mavie Cardi, and Jacopo Schettini Gherardini

Eugenics of Non-financial Information

The Double Genesis of Non-financial Information

The information—financial and otherwise—has a double genesis: it can be produced directly by firms and financial intermediaries which are active on the market; it can be produced by third parties, based on specific

The chapter is the result of a common work of the authors; nevertheless, it can be attributed to Mario La Torre paragraph 5.1, to Mavie Cardi paragraph 5.2, to Jacopo Schettini paragraph 5.3 and to Elena Olmedo paragraph 5.4.

M. La Torre (✉)
Università Di Roma La Sapienza, Rome, Italy
e-mail: mario.latorre@uniroma1.it

E. Elena
Universitat Jaume I, Castellón de la Plana, Spain
e-mail: eescrig@cofin.uji.es

N. Linciano et al. (eds.), *Information as a Driver of Sustainable Finance*, Palgrave Studies in Impact Finance,
https://doi.org/10.1007/978-3-030-93768-3_5

requests from operators, or for regulatory and supervisory purposes (see Chapter 3 for more details). The environmental, social, and governance (ESG) ratings, scoring, and opinions produced by information providers constitute the focus of this Chapter.

There is a dialectical relationship between "in-house information" and information prepared by third parties; the latter, in fact, feeds on the former, just as "in-house information" is itself designed to represent a useful database for external information providers. This is also true for the ESG world.

This dialectical relationship is the basis of the key question on which the analysis of this work revolves: how does this dialectic of "double genetics" affect the non-financial information of external providers?

The latter responds to the different needs of businesses and investors by following, as well explained in paragraph 5.3, different business models. In both cases, they are called upon to guarantee the market an independent judgment that cannot be ensured by information produced internally by companies.

As for the Reliability of Primary Information

The reliability of primary information made available by businesses is presupposed for the quality of information produced by third parties. Yet, in the field of non-financial information, this reliability cannot be taken for granted; the qualitative nature of many elements concerning non-financial information leads to discretionary assessments to a much greater extent than is the case for financial information. It comes as a logical result that the process of producing non-financial information by third parties must, in the first place, be supported by a robust assessment of the reliability of primary information produced by businesses. This inevitably

M. Cardi
Link Campus University, Rome, Italy
e-mail: m.cardi@unilink.it

J. Schettini Gherardini
Standard Ethics, London, UK
e-mail: jacopo.schettini@standardethics.eu

requires a tailor-made approach that may be badly combined with artificial intelligence models, increasingly used by information providers, to the detriment of expert analysts' assessments. The relative youth of the ESG culture within the business makes this dynamic even more relevant, especially if contextualized in smaller businesses, less equipped in terms of economic resources and know-how, in setting up consistent and reliable materiality analysis of ESG factors, as well as data processing. An unreliable database renders any evaluation flawed, even if based on correct processes and analysis, as discussed also in Chapter 3 where issues related to data gaps were mentioned; it is a paradox that the Stoics had clearly highlighted when they asserted that, starting from false premises, one could reach equally false conclusions with a perfect syllogism.

The literature analysis proposed in paragraph 5.2 highlights how quality control on the production of non-financial information is still negligible, and how the market is compensating for it with an overproduction of information, generally summarized in key performance indicators (KPIs) and worked by artificial intelligence systems. Literature also seems to assume the reliability of information as given when it explores relationships between ESG sustainability factors and company or portfolio performance, as well as between ESG factors and market reactions.

As for the Business Model Adopted for the Production of Non-financial Information

Once the reliability of primary information has been validated, the identification quality of the information itself must be valued in the context of the business model used: information from an external provider will have a different significance when requested by the investor or the issuer, since these have different perspectives and different objectives, as clarified in paragraph 5.3 and in paragraph 5.4. The distinction between scoring, ranking, rating, and opinion also finds reason in this assumption which, performing fundamentally distinct functions, take on diverse information and process it differently.

ESG scorings offer quantitative data to make decisions. These scorings are isolated data on KPIs that do not allow extracting more information to predict future events.

Meanwhile, ESG rankings offer classifications of companies according to their ESG performance, placing them in a certain order or grouping according to a specific system.

On their behalf, ratings focus on a company's default and business risk (Scholtens & Witteveen, 2021). ESG ratings use both quantitative and qualitative data for their evaluations and by means of complex formulas allow the prediction of future risks. Although each rating agency publishes assessments from different areas of analysis, the overall rating is the result of the discussion between groups of experts from the rating agency (rating committee). To that end, they can use public information (sustainability reports and information on corporate websites), as well as any other information that the client provides confidentially to carry out the evaluation of the company (Jackson et al., 2019). However, the quality of the analysis does not lie in the ability to obtain and measure the data, but rather in the ability to interpret and classify them effectively according to clearly defined criteria. The rating can be mandatory and paid for by the issuer or by an investor—to reduce the information asymmetry (Cui et al., 2018).

This distinction creates an insurmountable gap between rating on the one hand and scoring and opinion on the other. Yet, both literature and ESG world operators struggle to identify this demarcation, which in traditional finance is universally recognized. In literature, for example, comparative analyses are a la mode, tending to highlight differences in judgment between different information providers, which unite ESG ratings and ESG scoring in the sample observed, generating twice as much confusion.[1]

As for the Correct Use of Non-financial Information

As clarified in paragraph 5.4, the perspective and the need of the user regarding information is the basis of precision in the use of information itself. The debate on banks' non-financial disclosure can be a prime example of what has been mentioned. In the banking sector, the ESG issue is strongly focused on the perspective of environmental and social risks, and in particular on those relating to the intermediary's credit portfolio. From this perspective, it is essential that banks can count on ESG ratings more than scorings, since ratings are built according to a forward-looking approach based on indicators and algorithms capable of capturing exposure to risk. Thinking of measuring banks' ESG risk exposure using scoring—which, generally, produce information based on output-based

KPIs—would inevitably be misleading. From this perspective, hence the request of banking authorities—and the work of the European Banking Authority (EBA) in particular[2]—to include ESG factors in the risk appetite framework (RAF) of European banks will remain an unfinished operation until a real industry of ESG ratings will not be able to meet the demand for sustainability ratings in a widespread manner, to cover the numerous positions—also referring to smaller companies—contained in vast credit portfolios of banks.

As for the Significance of an ESG Rating Industry

It is in this context that paragraph 5.4 explains the need to speed up the process of setting up an ESG rating market that can allow, with the due freedom of rating provider modeling, reliability of primary information data, clarity of the business model, and of the production process, comparability of judgment between providers. The same scheduled supervisory obligations on banks and asset managers, provided for by the EBA and ESMA (European Securities and Markets Authority), aimed at managing ESG risks, will not be able to be concretely implemented without the aforementioned conditions being met.

The work of the European legislator, who followed the path outlined by the 2018 Action Plan on Financing Sustainable Growth (see Chapter 8), must find an equal pace of progress between various compliance provisions on sustainability and development of a financial reporting market for third parties, in its various ratings, scoring and opinion dimensions.

Third Parties' Non-financial Information: A Systematic Literature Review

The attempt to define the relationship between ethics and finance has led scholars to critically examine "*over thousands of years [...] some aspects of traditional analyses of financial ethics and economics, varying from Kautilya and Aristotle in the fourth century B.C., to Adam Smith in the eighteenth century*" *as Amartya Sen famously stated. A route, this one, in which the literature's interest in the set of the three main pillars of ethical finance - Environmental, Social, and Governance - lies relatively more recently. ESG issues reflect contemporary challenges and further open the analysis to "some interrelated problems of financial and business ethics dealing with the duties of the financial agents, such as managers and other administrators of financial institutions and business firms*" (Sen, 1991).

As will be discussed in the following paragraphs, the voluntary and partly discretionary nature of *ESG duties*[3] has had some implications over time for the methodological and strategic profiles of disclosure and related investment decisions.

However, it is worth noting that, in the European context, regulation has gradually introduced—for large public-interest companies—disclosure requirements on how companies operate and manage social and environmental challenges.

As was discussed in Chapter 4, Directive 2014/95/EU—Non-Financial Reporting Directive (NFRD)—sets out the rules on disclosure of non-financial and diverse information by certain large companies. The rules are supplemented by the guidelines published by the European Commission (June 2017) with specific reference to the disclosure of environmental and social information. Additional guidelines on climate-related information reporting were published by the Commission in June 2019.

The current NFRD reporting requirements have recently been subject to a proposal for amendment by the Commission, which adopted a proposal for a Corporate Sustainability Reporting Directive (CSRD) on 21 April 2021. The proposal expands the range of companies that must comply with the above-mentioned regulation, including all large companies and all companies listed on regulated markets. The proposal also requires that the information reported would be subject to audit and introduces more detailed reporting requirements in line with mandatory EU sustainability reporting standards.

Further regulatory schemes to be mentioned here concern governance profiles: on the subject, the Directive 2017/828—Shareholder Rights Directive II (SHRD II)—establishes specific requirements to encourage shareholders' long-term engagement. Environmental and social issues are considered particularly relevant. These requirements apply in relation, inter alia, to the identification of shareholders, transmission of information, transparency of institutional investors, asset managers and proxy advisors, remuneration of directors, and related party transactions.

Literature has investigated the role of ESG factors from different perspectives. For the purpose of the following paragraphs, three main strands of literature gave cause for reflection, covering three different aspects (Table 5.1):

Table 5.1 Selected contributions by area of investigation

Investigation areas	*Key elements*	*Texts*
ESG indexes and ratings	ESG ratings divergence is referred to three sources: scope divergence, measurement divergence and weight divergence; lack of commonality in characteristics, attributes, and standards used for the purpose of defining the E, S, and G components; lack of convergence of ESG measurement concepts. Disagreement has increased over time. Divergence between sustainability ratings gives rise to critical issues in terms of their comparability.	Berg et al. (2019), Billio et al. (2020), Dorfleitner et al. (2015), Chatterji et al. (2016) Christensen et al. (2019) Boffo et al. (2020)
Sustainability assessment process	Evolution of ESG analysis over time, ESG research has moved from data scarcity to abundance and now superabundance with artificial intelligence. ESG ratings agencies do not fully integrate sustainability principles into the corporate sustainability assessment process. ESG ratings sustainability performance: lack of alignment between "E" pillar and overall ESG scores, at least by some providers.	Kuh (2019), Escrig-Olmedo et al. (2019), Boffo et al. (2020)

(continued)

Table 5.1 (continued)

Investigation areas	*Key elements*	*Texts*
ESG Information and market reaction *ESG Information and shareholder value* *ESG Information and portfolio performance*	Existence of a predictive ability of ESG ratings with respect to future ESG news. Increased transparency aimed at highlighting ESG outcomes can reduce the cost of equity capital. Less negative market reaction around the institutional announcement of the ESG disclosure mandate in the EU for firms that already have high ESG disclosure. Stock markets reaction to events related to CSR, as a proxy for ESG performance: investors respond strongly negatively to negative events and weakly negatively to positive events. Changes in ESG ratings, unrelated to any sustainability news, shape investors' portfolio allocation decision and cause temporary pressure on the stock prices involved. Positive relationship between carbon risk and cost of debt: European financial markets take into account companies' exposure to carbon risk.	Serafeim and Yoon (2021) Sassen et al. (2016), Grewal et al. (2019), Kruger (2015) Latino et al. (2021) Pizzutilo et al. (2020) Naughton et al. (2019), Flammer (2013), Capelle-Blancard and Petit (2019) Dhaliwal et al. (2011), Ioannou and Serafeim (2015), Eccles et al. (2014), Dimson et al. (2015), Brogi and Lagasio (2019), Khan et al. (2016), Gibson et al. (2019) La Torre et al. (2020) Auer and Schuhmacher (2016), Bannier et al. (2019) Starks et al. (2017)

Investigation areas	*Key elements*	*Texts*
	Announcements of ESG activity generate positive abnormal returns and vice-versa.	
	Positive impact of ESG efforts and disclosure on firms' financial performance. Positive impact of ESG on profitability.	
	Good ESG practices have different effects in different business areas.	
	Divergence between ESG ratings affects stock returns.	
	Performance does not seem to be affected by the efforts in terms of ESG commitments.	
	Active selection of ESG securities does not provide a superior risk-adjusted return when compared to traditional passive strategies.	
	Portfolios with low ESG scores compensate for higher risk with higher returns.	
	Investors with longer horizons tilt their portfolios toward firms with high ESG profiles.	

- the role of ESG indexes in incorporating ESG information, with particular regard to the multidimensionality of ESG ratings;
- the relationship between sustainability principles and the corporate sustainability assessment process carried out by ESG rating providers;
- the impact of ESG information on the market.

ESG Indexes and Ratings

With reference to the first profile, in the light of the subsequent remarks in the following paragraphs, it is interesting to highlight that a large part of the literature agrees in underlining a certain divergence of ratings expressed on ESG issues by different providers: we will attempt to give a reading of the general roots of the problem in the following paragraphs, describing the simultaneous presence of different evaluation metrics reflecting different business models, methodological approaches and objectives leading to different ratings in quantitative terms.

Specifically, Berg et al. (2019) investigate the causes of such "Aggregate Confusion" and differences in ESG ratings across providers by analyzing the disagreement between ratings expressed by five major rating agencies. More specifically, the authors state that, in the dataset of five different ESG rating providers, the correlations between their ratings are on average 0.61, moving from 0.42 to 0.73, while, by comparison, the ratings of the major credit rating agencies are correlated at 0.99. The research suggests that ratings vary due to three essential reasons: namely, because of the divergence in scope, related to the selection of different sets of categories; because of the divergence in weight, related to the relative importance of categories in the calculation of the overall ESG score, and finally, above all, because of the divergence in measurement.

Similarly, Billio et al. (2020) analyze the ESG rating criteria used by nine major agencies, demonstrating a lack of commonality in characteristics, attributes, and standards used for the purpose of defining the E, S, and G components. Heterogeneity in the ESG arena leads to discrepancies among ESG evaluations. The authors add that the presence of alternative definitions of ESG also influence the identification of benchmarks used by the sustainable investment industry. Analysis of four representative ESG indexes, from four different providers, shows that the common constituents only overlap by 15% due to disagreement in the scores provided by the rating agencies; however, this limited overlap in ESG indexes dissipates the effect of ESG investor preferences on asset prices, to the point that, even for the ESG agreement portfolio, there is no impact on the financial performances.

Similarly, Dorfleitner et al. (2015) argue for a clear lack of convergence of ESG measurement concepts. Moreover, since it is unclear how one would assess their validity ex post, this has likely led to the persistence of disagreement among raters over time; moreover, recent evidence suggests that this disagreement has been increasing over time (Christensen et al., 2019). Chatterji et al. (2016) argue that the views of six major providers have fairly low correlations with each other. In general, the divergence between sustainability ratings, mentioned in several studies, gives rise to critical issues in terms of their comparability (Boffo et al., 2020). In addition, divergence between ESG ratings affects stock returns: according to Gibson et al. (2019), there is a negative correlation between the dispersion of ESG ratings and stock returns of a sample of S&P500 companies, hence a higher disagreement between ESG ratings results in an overvaluation of stocks, and thus lower overall returns.

Sustainability Assessment Process

Another strand of literature focuses on a more specific perspective: the evolution of the analysis of ESG factors over time and the inclusion of sustainability principles in the corporate sustainability assessment process.

Kuh (2019) provides a review of the evolution of ESG analysis over time, observing how ESG research has moved from data scarcity to abundance and now superabundance with artificial intelligence. According to the author's reconstruction, the age of scarcity (1970s–1990s) corresponds with the rise of socially responsible, or ethical, investing: portfolio screening was the primary means of portfolio construction. In the age of abundance (1990s–2010s), with the growing interest in corporate responsibility on the part of investors, companies came under pressure for more disclosure and transparency on ESG issues. This period is characterized by the development of ESG ratings derived from a focus on materiality in an industry context: in this era, ESG analysis is no longer focused on searching for scarce information, but rather on a systematic activity of selecting and structuring information from the growing volume of available data. In this way, qualitative information was reclassified into quantitative data, also in order to distinguish meaningful from inconsequential information. Finally, the period that the author describes as information superabundance (2010s–and beyond) finds limitations and opportunities arising from the integration of technology and artificial intelligence into ESG ratings.

Escrig-Olmedo et al. (2019) analyze how the criteria used by ESG rating providers in their assessment processes have evolved over the past 10 years. The authors also note that ESG ratings agencies do not fully integrate sustainability principles into the corporate sustainability assessment process, with particular reference to the balance between sustainability dimensions, intergenerational perspective, stakeholder approach, and life-cycle thinking.

With reference to sustainability outcomes, Boffo and Patalano (2020) focus on the specific importance of information related to the "E" pillar in light of the growing demand for environmentally sustainable products, also in view of the attention that central banks and financial supervisors are giving to these profiles. The authors, therefore, inquire into the sustainability performance of ESG ratings, finding that the lack of alignment of E and overall ESG scores, at least by some providers, indicates that high-ESG portfolios are not necessarily aligned with strong environmental performance. Despite encouraging progress, the report notes that some aspects of Pillar E within ESG rating methodologies suffer from a certain level of inconsistency, resulting in implications for the reliability of these ratings in achieving sustainable outcomes.

ESG Information and Market Reaction

Finally, a third strand of literature focuses on the effects of ESG information on the market. Within this strand of literature—which is extremely wide-ranging and which can be just mentioned here—we can distinguish between contributions that focus on market reactions to ESG information and studies that focus specifically on the relationship between ESG factors and corporate financial performance: specifically, as part of the latter, we analyze below the contributions that focus on the correlation between ESG information and shareholder value and on the impact of the same information on portfolio performance (see also Chapters 3 and 6).

On an overall level, several studies have examined how the market reacts to events of different nature related to ESG issues. For instance, Grewal et al. (2019) examined market reactions around the institutional announcement of the ESG disclosure mandate in the European Union and documented a less negative market reaction for firms that already have high ESG disclosure. Moreover, from this perspective, increased transparency aimed at highlighting ESG outcomes can reduce the cost of equity capital (Sassen et al., 2016). Krüger (2015) studies how stock markets react to positive and negative events related to corporate social responsibility (CSR), as a proxy for ESG performance. By applying the methodology of textual news analysis following a behavioral finance approach, the author shows that investors respond strongly negatively to negative events and weakly negatively to positive events. Latino et al. (2021) study the impact of ESG ratings on investor preferences and stock prices: in particular, the authors provide causal evidence that changes in ESG ratings, unrelated to any sustainability news, shape investors' portfolio allocation decision and cause temporary pressure on the stock prices involved; more precisely, investors who misinterpreted firms' sustainability rankings rebalance their portfolios by buying stocks they perceive as more ESG-oriented and sell firms they consider less sustainable. On a more general level, Serafeim and Yoon (2021) demonstrate the existence of predictive ability of ESG ratings with respect to future ESG news, highlighting the importance of this ESG outcome, whose usefulness, however, decreases in the presence of significant disagreement. The study uses data from three of the largest ESG rating providers with broad coverage (i.e., MSCI, Sustainalytics, and Thomson Reuters) to test whether the circulating consensus ESG rating (i.e., the average across the three) is able to predict future ESG news. The results show a strong positive predictive relationship between ESG ratings and future ESG news, although this decreases significantly in the presence of significant disagreement. In a previous study, Serafeim (2018) already argued that ESG criteria and public opinion are interconnected, evaluating their consequential effects. Specifically, the study showed that the valuation premium paid for companies with strong sustainability ratings has gradually increased over time, and this can be attributed to the momentum of positive public sentiment: public sentiment influences investors' views on the value of corporate sustainability assets and thus both the price paid for corporate sustainability and the investment returns of portfolios that consider ESG data.

With specific reference to the correlation between *ESG information and shareholder value,* a large part of literature (Flammer, 2013; Naughton et al., 2019) finds that announcements of ESG activity generate positive abnormal returns, while conversely, negative market reactions to negative ESG news are found (Capelle-Blancard & Petit, 2019).

A large majority of studies stress the positive impact of ESG efforts and disclosure on firms' financial performance (Dhaliwal et al., 2011; Ioannou & Serafeim, 2015). Eccles et al. (2014) follow a combined approach (using data from different sources) to identify high and low sustainability firms from a sample of 180 U.S. firms and find a significantly better operational and market performance by firms that meet the sustainability criteria. Similar results are identified with respect to the impact of environmental practices (Dimson et al., 2015). However, good ESG practices are also found to have different effects in different business areas, depending on the correlation between the single elements and the company's business (Khan et al., 2016). With particular reference to environmental risk, the study conducted by Pizzutilo et al. (2020) focuses on the relationship between one of its specific components, namely the so-called carbon risk, and the cost of debt. The analysis is carried out on a sample of companies listed in the Eurostoxx 600 index, to investigate the relationship between the carbon emissions of these companies and the cost of debt in the years 2010–2017. The results show a positive relationship between carbon risk and the cost of debt, highlighting that European financial markets take into account companies' exposure to carbon risk.

With particular regard to the banking sector, Brogi and Lagasio (2019), using MSCI ESG KLD STATS data, with reference to a large sample of U.S. listed companies from 2000 to 2016, provide evidence of the positive impact of ESG disclosure and company profitability, as measured by return on assets (ROA), noting this positive association especially in banks compared to industrial firms.

However, there are also studies that do not support the previous evidence: La Torre et al. (2020) analyze how ESG components affect the stock returns of firms included in the Eurostoxx50 index over the period 2010–2018 and find that performance does not seem to be affected by the efforts in terms of ESG commitments of the analyzed firms.

Finally, from an *ESG portfolio performance perspective*, many scholars analyze different portfolio strategies to understand how performance is affected (see Chapter 6 for additional discussion on this topic). Auer and Schuhmacher (2016) analyze the performance of ESG portfolios using Sustainalytics indicators with reference to different regions, US, EU, and Asia, and with an eight-year time horizon starting in 2004. The study considers different portfolio screens and finds that active selection of ESG securities does not provide a superior risk-adjusted return when compared to traditional passive strategies and in Europe in particular, performance can be negatively affected with reference to specific sectors. In contrast, the study carried out by Henke (2016) measures the financial impact of ESG-based screening on corporate bond portfolios: an analysis of the financial performance of 103 U.S. and European socially responsible bond funds (2001–2014) shows outperformance of the latter, directly related to ESG risk mitigation, compared to a corresponding sample of conventional funds.

Bannier et al. (2019) analyze ESG score portfolios in the U.S. and Europe, showing that portfolios with low ESG scores compensate for higher risk with higher returns. Starks et al. (2017) focus their analysis on the criterion of investors' time horizon, finding that investors with longer horizons tilt their portfolios toward firms with high ESG profiles.

Overall, the existing literature reveals largely mixed empirical evidence, often highlighting the difficulty of quantifying the true impact of ESG rating on portfolio performance. Factors influencing this inconclusiveness among others include issues with different providers and methodologies, as discussed in the following paragraphs.

Concerning the Business Model of Third Parties' Information Providers

Investor-Pay Model vs Applicant-Pay Model in Non-financial Recommendations

The rationale behind a certain pay model is strictly connected to the kind of financial services offered. As for financial advisory services (FAS) to institutional and retail investors—including buy/sell recommendations, analysis, and scoring of listed companies—providers that apply the *Investor-pay Business Model* are regulated in all major OECD countries.[4] FAS are often tailor-made services, which is why they are assessments and not necessarily based on the same analytical methodology. In fact, neither the comparability of the judgments, nor the obligation to publish are envisaged. According to the various national legislations, there are strong limits to the treatment of price-sensitive information. Or rather, the underlying rationale is that price-sensitive information cannot constitute a privileged assessment for one or a few. The crimes that can be derived from the various legislations also include stock manipulation. The documentation to which those who are engaged, directly or indirectly, by investors have to access is therefore standard documentation, such as public balance sheet data.

On the other hand, agencies that issue credit ratings are subject to other rules, obviously without interfering with the methodologies adopted by the agencies.[5] Their output is designed to be public and comparable, and it is the regulator itself that validates algorithms and methodologies and establishes publication rules. The methodology adopted by the rating agency is not changed at the request of the client, according to his priorities driving investment decisions. No recommendations or solicitations are proposed; the evaluations must be accessible to the whole market at the same time; the paying customers cannot be the investors, but are the companies interested in being rated; rating agencies adopt the so-called *Applicant-pay Model*. The rationale for why rating agencies should not be paid by investors is that they should remain neutral and independent from the market,[6] and not work for an individual investor: the ratings are disclosed through special channels or are read through the media. In this sense, the collection and management of information are also different from consulting and may include price-sensitive information. In fact—among other things—the data as such remains secret, but its processing is necessary for the rating agency to develop an "opinion" concerning risks. Opinions that are however intended for the entire market, not reserved for a single financial player. Moreover, in gathering information, the agency can count on the (contractually regulated) cooperation of its company clients (Applicants), which provide detailed and even confidential information, such as business plans or internal procedures.

In the context of traditional finance, these two types of business models—*Investor-pay Model* and *Applicant-pay Model*-, appear to be well developed, complementary and regulated. The first with the aim of generating assessments that investors use to achieve their investment objectives and outline their portfolio in terms of "priorities" and "opportunities". The second model focuses on measuring risk, which is also used by investors to outline their portfolio, but this time in terms of "riskiness". As is evident, these are two business models that differ, not only in the business and commercial organization, but also in the management of conflicts of interests, in the methodologies of analysis and, as we have seen, in the collection and management of information.

This division of roles, briefly summarized here, it is still not perceived in the ESG sector.

In fact, according to a recent study published by ESMA,[7] the business sector of the ESG evaluations is mainly made up of consultants and data providers who work on the side of investors, while on the side of the *Applicant-pay Business Model*, as indicated in a small note in the *Final Report of the High-Level Expert Group on Sustainable Finance* of the European Commission,[8] only one agency—Standard Ethics—is active, which has voluntarily self-regulated on the model of credit rating agencies.[9]

The reason for the development of the ESG valuation market, which is so asymmetrical and so investor-focused, has two basic rationales: an entrepreneurial and a methodological one. Both are related to the notion of voluntariness.

It has always seemed clear-cut that actions in the area of ESG should be voluntary and not a legal obligation. As a consequence, it has not been possible to make ESG ratings mandatory for issuers. The absence of mandatory status weakened this business case from the outset. It was assumed that no issuer would voluntarily bear the cost of undergoing a non-mandatory rating such as an ESG rating.

The second reason is more conceptual: contrary to general non-financial information, the non-mandatory nature of ESG ratings may be explained recurring to the *Stakeholder Theory* which indirectly supports the arbitrariness of ESG rating policies, leaving them to a materiality analysis and a relationship with stakeholders. This vision is consistent with a part of liberal thinking that places corporate freedom at the heart of its concerns and, therefore, places an ESG strategy in a functional position subordinate to the interests of the company. We are faced with a vision that we would call "Ptolemaic", in which the economic entity that implements socially or environmentally responsible policies is at the center. It is a doctrine that is all in all "convenient" because the company—and the investor—enjoys great arbitrariness in the choice of actions to be taken and objectives to be selected.

This also explains the marketing of indices with the most diverse characteristics, ranking or scoring models based on very different principles, and systems for collecting data and feeding databases to meet the most diverse needs. One can understand how investors' demand for companies specializing in ESG consultancy services has grown exponentially. It is a process in which sociologists, environmentalists, and energy experts enter the field. A process that has led to reporting standards, such as Global Reporting Initiative (GRI).

On the other hand, the scale of this arbitrariness, which is so useful for business growth for investors, poses a methodological limit to the construction of ratings—if one attributes to this term the definition given initially. It is in fact evident that, if we talk about comparisons with some degree of scientificity, we cannot holistically compare companies, even in the same sector and of the same size, if they choose to implement different CSR policies. If they choose to compete on different terrains. If they also choose different targets and timing. At that point, a holistic comparison could only take place on a value level, because it would be expressed on ethical or subjective choices, on materiality matrices that represent legitimate, even if divergent, interests. The remaining alternative is to compare companies on specific issues, which are selected, limited, measured, and weighed. Which is what is happening now.

From a market observatory, it is therefore possible to attribute—not only *to commercial reasons but also to methodological limits on comparability*, the failure to develop services dedicated to "Applicants".

The same asymmetry can be found in academic literature. A literature absorbed by the *Investor-pay Model* paradigm: yesterday focused on explaining the correlations between financial performance and ESG issues, today concentrated on finding the inhomogeneity between evaluations. A literature that is not very committed to describing the general roots of the problem.

This lack of in-depth analysis is probably also the basis of the reason why the request for full comparability of assessments has so far been addressed to the wrong interlocutors: asset managers. Indeed, if we wanted to achieve holistic comparability in this business sector, in which investors pay, then we would have to standardize all the ESG investment objectives of all investors. Then impose the same point of view on asset managers. Thus, we would have a uniform ranking and scoring. If this were the case, some questions would arise: who would be responsible for defining the investment objectives, and therefore the structure of the ESG portfolios if we were to remove this faculty from asset managers? Would it be up to investor lobbies, environmental or human rights associations, the third sector, a political body, one or more nation-states? But the really most important and crucial question would be around the relationship between saver and manager: who would be responsible for the legal and commercial responsibility of an ESG portfolio? Who is responsible for the economic performance of an ESG investment fund?

From how the discussion in Brussels is evolving, it seems that for some, the taxonomy of the European Commission itself must fulfill this function. But if this is the case, one might wonder whether one of the fundamental principles of economic freedom is not being violated, according to the fact that eligible sustainable activities are restricted to a specific set suggested by the taxonomy. A principle that, moreover, is fundamental to the sustainability that international organizations themselves have defined as such. This approach is a bit like taking the crocodile by the nose: it is impractical and seems to be based on the idea that there is no other business model than the *Investor-Pay Model*, and that the same methodology can be used to holistically measure risks or to measure investment opportunities.

The Case of Standard Ethics

In this context, an interesting case to be empirically examined in the market today is that of Standard Ethics—since 2004 operating under the name Aei-Standard Ethics GEIE, and since 2014 as Standard Ethics Ltd.

Standard Ethics developed building on three basic assumptions aimed at offering comparable, uniform, investor-independent ESG ratings, modeled on credit ratings, should be based on three basic assumptions. These three assumptions can be found in the rules that national and international financial authorities have been applying to credit rating agencies.

The first of these three assumptions would be to serve a type of client other than investors, that of "applicants", following rules of exclusivity to avoid conflicts of interest. This choice, however, should deal with the absence of the obligation for companies to have an ESG rating, therefore with all the entrepreneurial risks involved. This is the first unknown factor.

The second assumption would be to be able to compare companies in a uniform way with respect to a single point of view. A fixed point is represented for the analyses of credit ratings by the notion of default.

The third assumption, achievable thanks to the first two, would be to implement analysis methods focused on risks, and not on investment objectives and opportunities.

Regarding the first condition, the company bases its business—at the statutory level—on the sale of solicited ratings, it has no economic relations with investors. This issue was resolved at the base through a very detailed Article of Association from which specific procedures derive.

The company organization of Standard Ethics has a very particular morphology, graphically reminiscent of that of a flower, and in fact it is described as a "Flower Pattern" because it divides the areas of activity functional to the marketing of the ratings and the underlying analyses into "petals". Areas separated by Chinese walls.[10]

With reference to the second condition, that is to make its ratings comparable by adopting an absolute and not relative-based measurement scale, the company has identified a fixed point to which to standardize its measurements. To achieve this objective, only the voluntary indications of sustainability coming from the UN, the OECD and the European Union were placed at the center of each evaluation, excluding any ethical option and without attributing weight to the materiality analysis of the company.

In 2002,[11] and subsequently in other works from 2005[12] onwards, analysts at Standard Ethics introduced a difference between the model referring to the *Stakeholder Theory* and a second model that could be implemented by companies, which would be based exclusively on voluntary indications coming from supranational organizations. From a methodological point of view, it was thought that this approach could provide the cornerstone around which to build an algorithm and a standard measure good for a (holistic) sustainability rating.

In practice, a conceptual separation has been theorized between the notion of Corporate Social Responsibility (CSR) and that of Sustainability. In the latter case, following the definition of the Brundtland Report (planet and new generations), it was possible to motivate companies and investors as to why, according to Standard Ethics, it was out of their hands to determine what should, or should not, be considered sustainable on a global level. How the direction of the so-called ethical compass was no longer in the hands of the company and its stakeholders, but automatically aimed at an ideal north indicated by international objectives and guidelines.[13]

Companies observed by Standard Ethics from this perspective are not restricted in their management freedom, but they know that the playing field within which they are measured for the rating is only that defined by global objectives and bodies. There is therefore an indirect limitation to the arbitrariness of company ESG policies, without this preventing the company from making different choices.

The third area, that related to the methodology of analysis and data collection (appropriate to the study of risks), as well as the type of output, is all issues resolved in a similar way by the model of credit rating agencies.

Certainly, Standard Ethics is very far from the modus operandi of the *Investor-pay business Model.*

Data is collected with the full cooperation of the client. No questionnaires, or artificial intelligence software or programs, are used, but the data is managed by analysts. There is a contractual relationship that includes a *Non-Disclosure Agreement*, dealing with aspects that would not under any circumstances be made public or available to third parties if they were not already. Information is segregated on a client-by-client basis because a database is not maintained for consultancy purposes.

The evaluation methodology is not tailor-made and cannot be modified under any circumstances according to the client's request. No ranking or scoring built around a client's requirements is offered.

Given that the declared objective of the Standard Ethics rating is to measure the distance between an issuer and the voluntary international indications on sustainability, Standard Ethics find itself, as with credit ratings, faced with a measure on an absolute basis, not a relative one. For Standard Ethics, it is neither necessary, nor functional, to build homogeneous sets of companies of the same sector and size in order to make comparisons within the sets. Nor is it functional or useful to divide and "cluster" the company by homogeneous topics in order to measure topic by topic through KPIs, and then measure performance against some median.

Data collection appears to be poorer and more focused than a KPI analysis. The Standard Ethics observation is based on guidelines consisting of approximately 220 questions that aim to assess whether the company is aligned or not. Basically, each question does not give a score, but analyses the presence or absence of a risk. They are therefore "markers" designed to indicate, or exclude, the possibility of a problem. The aim is to map "conspicuous" points, and to mark decisive elements. This is why the final measure is not the algebraic sum of the individual KPIs, but the result of an algorithmic formula that the agency publishes (at least in its general form).

The Standard Ethics proprietary algorithm is based on five variables (plus "k"). The basic principle is based on two steps: the first is to verify certain conditions that are "necessary", but not "sufficient", to be defined as sustainable (fair competition; fair taxation; economic democracy; shareholders' rights, etc.); the second step is to more properly assess ESS policies. The lack of "necessary" conditions devalue any corporate commitment to EGS policies.

The reporting proposed by Standard Ethics—the *Final Report*—is, on the whole, in line with the output of a credit rating agency, since it does not provide quantitative data for investment choices, but aims to explain and justify the reason for a rating—a valuation that is summarized through a class of nine letter grades, from EEE to F. The Standard Ethics client, like any other "Applicant", wishes to be positioned and measured in relation to other rated companies, not in overall comparison, but absolute terms, i.e., in relation to international guidelines. Moreover, the Applicant wishes this measurement to be explained and commented on to the market, which, let us not forget, is the real end user of the rating.

It should be noted, once again, that Standard Ethics does not have an investor as a client, as is the case with Data Providers. The Data Provider does not know in advance what the point of view of an asset manager might be, so it is equipped to have a well-stocked database, with homogeneous numbers for each topic, from the amount of CO2 emissions to the number of accidents at work. The Data Provider will be interested in multiplying the KPIs to cover as many topics as possible; in building clusters for each topic; in calculating medians to assess the position of the individual issuer on the various topics. Its business is to sell data that can be aggregated according to various needs, leaving it up to the final decision-maker, such as the investor or the manager of an index, to weigh the KPIs arbitrarily. The analysis derived from the *Investor-pay business Model* is, therefore, a relative analysis by comparison. Aggregating this data may say something about the relationship between a single company and the others that make up its whole, but it does not add much in the way of absolute risk. As a mathematician like Bruno De Finetti would have said, the calculation is correct but says nothing. It is this gap that should be covered by ratings.

From a purely economic and commercial point of view, regarding the possibility of developing the *Applicant-pay Model*, Standard Ethics reports that it has been using the *Applicant-pay Model* in Italy as a test market since 2014.

From the current evidence available, the quality and quantity of large listed issuers that have applied for a solicited rating appear significant. This suggests that, even in the absence of mandatory ESG ratings, there is some proportion of companies that still wish to be rated by a third party. A body that is independent of investors and adopts a comparable rating. It is, of course, not clear how large this market is.

Discussion: The Road to the Future Structure of the Non-financial Information Market

As has been shown in this chapter, the integration of ESG factors in the assessment carried out by the financial market and, especially, by the ESG indexes and rating agencies, is a subject of study that needs to be addressed, both from a practical and academic perspective. It seems necessary to be able to face, on the one hand, the demand for information about ESG issues from different stakeholders and, on the other hand, the challenges of measuring and evaluating corporate sustainability.

Different stakeholders demand information on ESG aspects from the financial market for the decision-making process: investors (mainly institutional investors and asset managers), shareholders, regulators, or companies (Diez-Cañamero et al., 2020; Escrig-Olmedo et al., 2019).

Focusing on investors, they increasingly use non-financial, environmental, and social factors to evaluate a company's performance and make investment decisions (Busch et al., 2016) while building a vision of the long-term value and looking for a formal framework to assess and communicate intangible value.

However, What Information Do Investors Actually Receive from ESG Rating Providers for the Decision-Making Process?

Sustainability-conscious investors find a wide range of ESG data in the financial market. These ESG providers, by means of public information disseminated by the company or by alternative company data sources, evaluate corporate sustainability and provide the market with scorings, rankings, and ratings so that investors make their investment decisions (Fig. 5.1). However, the diversity of provider business models, with their complex and non-transparent methodologies and their different results, can make it difficult for investors to make decisions.

As has been evidenced in the literature review, some authors (e.g., Berg et al., 2019; Billio et al., 2020; Chatterji et al., 2016) highlight the divergences of ESG ratings. Divergences related to thematic areas or attributes, with the weighting and aggregation of the information and with the measurement and type of indicators (Berg et al., 2019). These divergences can be explained because the concept of corporate sustainability is a multidimensional concept and the process of adding qualitative and quantitative information on different ESG aspects can present certain

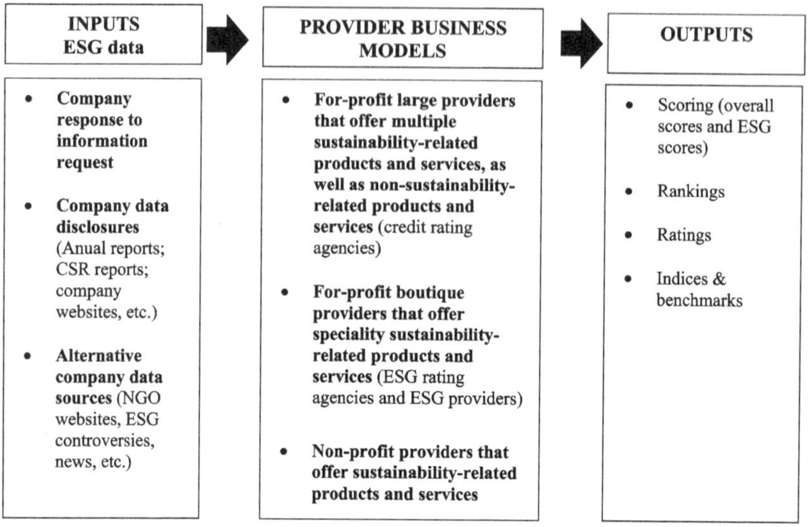

Fig. 5.1 Input/output/business models of ESG rating providers (*Source* Authors' elaboration based on European Commission [2020])

difficulties for rating agencies, such as the trade-offs among criteria (Escrig-Olmedo et al., 2017a).

These discrepancies and difficulties in the assessment process mean that sometimes there is no correlation between the ESG ratings given to a company by different ESG rating providers. These differences are much more frequent than in credit ratings. Therefore, they represent a major obstacle to the adoption of sustainable criteria by investors, since the information is incomparable and difficult to interpret (Juravle & Lewis, 2008). Moreover, these discrepancies also make it difficult for companies to identify the right strategy to improve their ESG performance. For this reason, it seems necessary to improve the quality, quantity, and accessibility of information. Indeed, in view of such divergences, some institutional investors choose to develop their own ESG rating systems, rather than use the information provided by ESG information providers.

It should also be noted that the abundance of information (Kuh, 2019), together with the evolution of the analysis ESG criteria over time do not always consider the Principles of Sustainability (Escrig-Olmedo et al., 2019)—that is, sustainability dimensions (environmental, social and economic), intergenerational perspective (managing the future and the current needs), stakeholder approach (stakeholder preferences), and life-cycle thinking (managing the impacts)—has been an added difficulty when evaluating corporate sustainability and hinders the process of selecting the investment universe.

Finally, there is a large amount of information on sustainability in the market, but with little clarity on how much of all this is related to credit. From the academic field, an attempt has been made to shed light on this issue, as evidenced by the review of the literature in this chapter, trying to advance in three fundamental aspects: ESG information and market reaction; ESG information and shareholder value; and ESG and portfolio performance.

However, is the Information that Investors Are Demanding to Make Their Investment Decisions in the Current Context of Global Risks?

ESG rating providers are an important intermediary between companies and investors. So much so that, institutional investors are increasing the demands to evaluate company performance using ESG factors, at a time when the rules for capital markets are being rewritten, and the economic and social impacts of the COVID-19 pandemic continue to unfold in a global context. They do not seek a large amount of data for their decision-making process, but rather efficient interpretation and classification according to clearly defined ESG criteria to take their investment decisions according to their investment objectives and preferences.

As already mentioned in the previous sections (see also Chapter 7), it is not possible to consider all sustainability-conscious investors as a homogeneous group, as they have different objectives. Escrig-Olmedo et al. (2017b) define in their work 3 types of investors:

- conventional investor, who invests with the main objective of maximizing financial returns. The financial and governance aspects of organizations are very important to this type of investor, while environmental and social concerns are less important;
- strong socially responsible investor, who invests based on their social interests and personal values. They perceive ESG aspects as very important in the investment decision-making process without ignoring financial criteria, although these are not a priority;

- instrumental socially responsible investor, who prioritizes good financial performance without neglecting ESG issues. Therefore, this investor considers financial aspects as very important in the investment process and ESG criteria as important.

Standardizing all ESG investment objectives for all investors is a difficult task and the financial market has an important challenge: integrating the expectations of the stakeholders in the measurement models of corporate sustainability (Escrig-Olmedo et al., 2017a).

Furthermore, ESG aspects are also a priority for many companies, affecting their management strategies, compensation policies, investment decisions, or public relations activities. In this context, companies need the proper information to identify ESG areas for improvement to make future management decisions that align with their organization's sustainability goal, industry goals, and stakeholder's goals.

In this regard, the importance of ESG rating providers is increasing, on the one hand, to provide comparable, accurate, and periodic information for investment decision-making and, on the other hand, to encourage responsible policies in companies. In this context, companies that do not meet investors' expectations of environmental, social, and governance factors run the risk of losing participation in capital markets.

From a regulatory perspective, the demands for investors and asset managers on ESG information are also increasing. In this regard, the already mentioned *Action Plan for Financing Sustainable Growth* was presented by the European Commission in March 2018 with the aim of reorienting capital flows toward a more sustainable economy, foster transparency and long-termism, and integrate sustainability into risk management (see Chapter 8). The plan aims to "*explicitly require institutional investors and asset managers to integrate sustainability considerations in the investment decision-making process*". As a result, requests for information on ESG matters from listed companies are becoming more frequent and more intense. In this context, ESG information analysts appear to be a linchpin.

ESG ratings and information provider agencies are unregulated and unsupervised, unlike counterparties focused only on financial reporting. This can lead to increased risks of greenwashing and capital misallocation. However, efforts are being made to generate more transparency and standardization in the industry. Beyond updating its guidelines on credit rating actions to improve transparency when sustainability factors have been taken into account, on 28 January 2021, the European Securities and Markets Authority (ESMA) published a letter in which it proposed to the European Commission to develop a European regulation for sustainability rating agencies, considering that ESG ratings and the assessment tools for these securities are essential for the implementation of the ESG regulation (see also Chapter 10). Specifically, ESMA[14] proposes:

- a common legal definition of what is meant by an ESG rating, capturing the broad spectrum of assessment tools available on the market;
- that any legal entity whose activity includes the issuing of these ESG ratings and assessments should be registered and supervised by a public authority;
- specific products requirements applicable to the ESG ratings and assessments provided by entities;
- organizational and conflict of interest management requirements that reflect their growing importance in sustainable finance.

Therefore, the ESMA focuses on an issue of vital importance for investors, such as the verification of ESG information and the need to make more efforts to integrate processes that ensure that information. Some ESG information providers claim that they verify the information through external sources, such as news or databases that send alerts when irregular behavior occurs—in sustainability—or other third-party sources that do not detail. Nevertheless, there is no evidence that they follow a process to ensure that information is in a standardized way. The challenge will be to define a standardized supervision system.

On the other hand, the European Union Action Plan for Financing Sustainable Growth (European Commission, 2018) requires that ESMA fosters the incorporation of sustainability risks in the investment decision processes, being credit rating agencies crucial in this process. ESMA should pay special attention to how the credit rating agencies are integrating corporate sustainability in their assessment process, as highlighted in Action 6 of the European Commission's Action Plan.[15]

With a view to fostering Action 6 "Better integrating sustainability in ratings and market research", in November 2020, as part of the European Commission's Action Plan on Sustainable Finance, the "Study on sustainability-related ratings, data and research" (European Commission, 2020) was published. It is an in-depth study that addressed, among other issues, the role, use, and quality of sustainability-related products that are provided by ESG information providers and market, the main players, data sourcing, transparency of methodologies and potential shortcomings in the sustainability assessment market. The study shows that ESG information providers measure, weigh, and score corporate sustainability risks and performance in different ways, being necessary for greater transparency. To strengthen disclosure on how ESG factors are being considered, in July 2019, ESMA updated its "Guidelines on disclosure requirements for credit ratings" (ESMA, 2019) and has started checking how credit rating agencies apply these new guidelines.

On 3rd February 2022 ESMA publishes a Call for Evidence on ESG ratings. The aim is to gather information on the market structure for ESG rating providers in the European Union (EU). In detail, the Call for Evidence's purpose is to develop a picture of the size, structure, resourcing, revenues and product offerings of the different ESG rating providers operating in the EU (ESMA will share the feedback received with the European Commission).

Complementary to ESMA Call for evidence, on 4th April 2022, the European Commission launched a public consultation aimed at exploring the dynamics of ESG rating maket and gathering evidence on the interplay between market players (investors, asset managers, benchmark administrators and companies subject to ratings). In addition the consultation focuses on possible shortcomings on the consideration of sustainability risks by Credit Rating Agencies (the consultation period closed on 10th June 2022).[16]

Therefore, disclosure and transparency requirements seem essential to integrate ESG risk analysis into investment decision-making processes. In this context, the best international response to face these risks, seeking sustainable development, is the *2030 Agenda for Sustainable Development* (the so-called Sustainable Development Goals—SDGs, UN, 2015).

Therefore, the generation of information on the main global sustainability risks and the contribution to the fulfillment of the SDGs is a challenge for companies. But, also for the financial market that seeks to attract and respond to new investors more aware of environmental and social issues and eager to contribute to sustainable development and climate change mitigation (Drempetic et al., 2019).

For this reason, ESG rating providers must work to harmonize the assessment criteria and to define the KPIs that are useful for measuring corporate sustainability and for identifying and managing future risks. To do this, an important aspect must be considered, such as the determination of the relationship between credit risk and ESG risk. It is a crucial aspect because it will make it possible to clarify the different probabilities of default associated with the different ESG ratings granted.

To sum up, ESG information providers should make efforts to integrate the new challenges of sustainability measurement and the expectations of the different stakeholders in their sustainability assessment models. Some of these market players are working toward it, as is the case study in this chapter of Standard Ethics, that implement analysis methods focused on risks, and not on investment objectives and opportunities, and has voluntarily self-regulated on the model of credit rating agencies.

Notes

1. For the same reason, it is found that numerous non-financial statements prepared by companies report the ESG judgments obtained by the various information providers without a clear distinction considering the real nature of the judgment itself.
2. Art. 98 (8) of the Capital Requirements Directive (CRDV) and art. 35 of the Investment Firms Directive (IFD) mandate the EBA to develop uniform definition of ESG risks and criteria for their assessment.
3. Voluntariness has gradually been overcome, through non-financial disclosure obligation (NFRD) as for companies and through SHRD II for asset managers.
4. In Europe, the legislative environment for financial advisory services is pictured by the so called MIFID II (2014/65/EU—Market in Financial Instruments Directive) and MIFIR (EU 600/2014—Market in Financial Instruments Regulation).
5. At the European level, the current regulatory framework on CRAs consists of a regulation (Regulation No. 462/2013) and a directive (Directive 2013/14/EU). Regulation (EC) No. 1060/2009 regulates also the issuance of ratings clarifying that the core of the rating is based on the fact that it should have as its final destination, not its principal (the applicant), but the overall market, simultaneously, in the same way, with the same data and the same access opportunities; this being not exactly the philosophy behind the consultancy business.

6. However, the Applicant-pay model still raises concerns about potential conflicts of interests and independency of the agency of the paying issuer.
7. ESMA Report on Trends, Risks and Vulnerabilities No. 1, 2021. ESMA50-165-1524. European Securities and Markets Authority, Paris. 17 March 2021.
8. Final Report 2018 by the High-Level Expert Group on Sustainable Finance. European Commission, Bruxelles, 31 gen 2018. p. 77.
9. www.standadethics.eu.
10. There are four main 'petals'. Each "petal" brings together the main areas of activity.

    A first "petal" brings together the functions intended to manage the solicited rating in the field, which the company calls"SOL". This business area includes several work units to manage a rather complex pipeline that starts with client relations, management and protection of the information received, bearing in mind that it must under no circumstances be shared with third parties or the market. The process ends with the drafting of a summary document, the proposal and approval of the rating with the involvement of a committee. A third petal is the monitoring activities that Standard Ethics also issues Unsolicited Ratings (UNSOL) to those constituents of its indices who are not clients (second "petal"). It has to be taken into account that the agency makes public all details about its indices, including the constituents and weights since 2004. The indices are in fact public and are not subject to commercial activity.

    A third 'petal' is the communication activity. In this case, the principles adopted by the credit rating agencies are followed, in principle, so that the ratings issued are communicated to the entire market simultaneously through agencies. The "fourth" petal is the Front Office, which includes the commercial activity. At the centre of the four 'petals' is a shared area which brings together those responsible for the individual sectors, without the rest of the organisation being able to cross the boundaries of confidentiality.
11. Schettini and Schettini (edited by), (2002).
12. Schettini (2005).
13. The first case, according to Standard Ethics, of the idea of sustainability is found in the OECD Guidelines for Multinational Enterprises of 1976. This notion was then brilliantly summarised in the 1987 Brundtland Report. This was followed by the United Nations Global Compact in 1999 and the European Commission's 2001 Green Paper on CSR—a document that was a little cautious compared to the OECD Guidelines. In which the European Commission, then led by Romano Prodi, had an ambiguous position. Soon, as is well known, the role and policies of international organisations and the European Commission changed radically.

14. For more information please read the ESMA letter: https://www.esma.europa.eu/sites/default/files/library/esma30-379-423_esma_letter_to_ec_on_esg_ratings.pdf.
15. Action 6: Better integrating sustainability in ratings and market research.

    1. Starting in Q2 2018, the Commission will engage with all relevant stakeholders to explore the merits of amending the Credit Rating Agency Regulation to mandate credit rating agencies to explicitly integrate sustainability factors into their assessments in a proportionate way to preserve market access for smaller players. Commission services will report on the progress made on this by Q3 2019.
    2. The Commission invites ESMA to: (i) assess current practices in the credit rating market by Q2 2019, analysing the extent to which environmental, social and governance considerations are taken into account; and (ii) include environmental and social sustainability information in its guidelines on disclosure for credit rating agencies by Q2 2019 and consider additional guidelines or measures, where necessary.
    3. The Commission will carry out a comprehensive study on sustainability ratings and research by Q2 2019. It will analyse methodologies and explore aspects like the market structure of sustainability ratings and market research services, the depth and breadth of sustainability research assessments and scoring, and the independence of those research/scoring providers. The study will also explore possible measures to encourage sustainability ratings and market research.

16. The European Conmission consultation uses the term ESG ratings based on the definition provided in the International Organization of Securities Commissions' (IOSCO) final report on environmental, social and governance (ESG) ratings and data products providers (21 November 2021), available at the following link: https://www.iosco.org/publications/?subsection=public_reports.

References

Auer, B. R., & Schuhmacher, F. (2016). Do socially (ir)responsible investments pay? New evidence from international ESG data. *The Quarterly Review of Economics and Finance*, *59*, 51–62.

Bannier, C., Bofinger, Y., & Rock, B. (2019). *Doing safe by doing good: ESG investing and corporate social responsibility in the U.S. and Europe* (CFS Working Paper Series 621).

Berg, F., Koelbel, J. F., & Rigobon, R. (2019). *Aggregate confusion: The divergence in ESG ratings* (MIT Sloan Research Papers 5822-19).

Billio, M., Hristova, I., Latino, C., & Pelizzon, L. (2020). *Inside the ESG ratings: (Dis)agreement and performance* (SAFE Working Paper 284)

Boffo, R., Marshall, C., & Patalano, R. (2020). *ESG investing: Environmental pillar scoring and reporting*. OECD Paris

Boffo, R., & Patalano, R. (2020). *ESG investing: Practices, progress and challenges*. OECD Paris.

Brogi, M., & Lagasio, V. (2019). Environmental, social, and governance and company profitability: Are financial intermediaries different? *Corporate Social Responsibility and Environmental Management, 26*(3), 576–587.

Busch, T., Bauer, R., & Orlitzky, M. (2016). Sustainable development and financial markets: Old paths and new avenues. *Business & Society, 55*(3), 303–329.

Capelle-Blancard, G., & Petit, A. (2019). Every little helps? ESG news and stock market reaction. *Journal of Business Ethics, 157*(2), 543–565.

Chatterji, K., Durand, R., Levine, D. I., & Touboul S. (2016). Do ratings of firms converge? Implications for managers, investors and strategy researchers. *Strategic Management Journal, 37*(8), 1597–1614.

Christensen, D., Serafeim, G., & Sikochi, A. (2019). *Why is corporate virtue in the eye of the beholder? The case of ESG ratings* (Working paper).

Cui, J., Jo, H., & Na, H. (2018). Does corporate social responsibility affect information asymmetry? *Journal of Business Ethics, 148*(3), 549–572.

Dhaliwal, D. S.,Li, O. Z., Tsang, A., Yang, Y. G. (2011). Voluntary nonfinancial disclosure and the cost of equity capital: The initiation of corporate social responsibility reporting. *The Accounting Review, 86*(1), 59–100.

Diez-Cañamero, B., Bishara, T., Otegi-Olaso, J. R., Minguez, R., & Fernández, J. M. (2020). Measurement of corporate social responsibility: A review of corporate sustainability indexes, rankings and ratings. *Sustainability, 12*(5), 2153.

Dimson, E., Karakaş, O., & Li, X. (2015). Active ownership. *The Review of Financial Studies, 28*(12), 3225–3268.

Dorfleitner, G., Halbritter, G., & Nguyen, G. (2015). Measuring the level and risk of corporate responsibility. An empirical comparison of different ESG rating approaches. *Journal of Asset Management, 16*(7), 450–466.

Drempetic, S., Klein, C., & Zwergel, B. (2019). The influence of firm size on the ESG score: Corporate sustainability ratings under review. *Journal of Business Ethics, 167*(2), 1–28.

Eccles, R. G., Ioannou, I., & Serafeim, G. (2014). The impact of corporate sustainability on organizational processes and performance. *Management Science, 60*(11), 2835–2857.

Escrig-Olmedo, E., Fernandez-Izquierdo, M. Á., Ferrero-Ferrero. I., Rivera-Lirio, J. M. & Muñoz-Torres, M. J. (2019). Rating the raters: Evaluating how

ESG rating agencies integrate sustainability principles. *Sustainability, 11*(3), 915.
Escrig-Olmedo, E., Muñoz-Torres, M. J., Fernández-Izquierdo, M. Á., & Rivera-Lirio, J. M. (2017a). Measuring corporate environmental performance: A methodology for sustainable development. *Business Strategy and the Environment, 26*(2), 142–162.
Escrig-Olmedo, E., Rivera-Lirio, J. M., Muñoz-Torres, M. J., Fernández-Izquierdo, M. Á. (2017b). Integrating multiple ESG investors' preferences into sustainable investment: A fuzzy multicriteria methodological approach. *Journal of Cleaner Production, 162,* 1334–1345.
ESMA (European Securities and Markets Authority). (2019). *Final report—Guidelines on disclosure requirements applicable to credit ratings*. ESMA 33–9–320. https://www.esma.europa.eu/sites/default/files/library/esma33-9-320_final_report_guidelines_on_disclosure_requirements_applicable_to_credit_rating_agencies.pdf
European Commission. (2018). *Action plan: Financing sustainable growth.* https://eur-lex.europa.eu/legal-content/EN/TXT/HTML/?uri=CELEX:52018DC0097&from=EN
European Commission. (2020). *Study on sustainability-related ratings, data and research.* European Commission. https://op.europa.eu/en/publication-detail/-/publication/d7d85036-509c-11eb-b59f-01aa75ed71a1/language-en/format-PDF/source-183474104
Flammer, C. (2013). Corporate social responsibility and shareholder reaction: The environmental awareness of investors. *Academy of Management Journal, 56*(3), 758–781.
Gibson, R., Krueger, P., & Schmidt, P. S. (2019). *ESG rating disagreement and stock returns* (Swiss Finance Institute Research Papers 19-67).
Grewal, J., Riedl, E. J., & Serafeim, G. (2019). Market reaction to mandatory nonfinancial disclosure. *Management Science, 65*(7), 3061–3084.
Henke, H. M. (2016). The effect of social screening on bond mutual fund performance. *Journal of Banking & Finance, 67,* 69–84.
Ioannou, I., & Serafeim, G. (2015). The impact of corporate social responsibility on investment recommendations: Analysts' perceptions and shifting institutional logics. *Strategic Management Journal, 36*(7), 1053–1081.
Jackson, G., Bartosch, J., & Avetisyan, E. (2019). Mandatory non-fnancial disclosure and its influence on CSR: An international comparison. *Journal of Business Ethics, 162*(2), 323–342.
Juravle, C., & Lewis, A. (2008). Identifying impediments to SRI in Europe: A review of the practitioner and academic literature. *Business Ethics: A European Review, 17*(3), 285–310.
Khan, M., Serafeim, G., & Yoon, A. (2016). Corporate sustainability: First evidence on materiality. *The Accounting Review, 91*(6), 1697–1724.

Krüger, P. (2015). Corporate goodness and shareholder wealth. *Journal of Financial Economics, 115*(2), 304–329.

Kuh, T. (2019). *ESG research in the information age*. Truevalue Labs

La Torre, M., Mango, F., Cafaro, A., & Leo, S. (2020). Does the ESG index affect stock return? Evidence from the Eurostoxx50. *Sustainability 12*(16), 6387.

Latino, C., Pelizzon, L., & Rzeźnik, A. (2021). *The power of ESG ratings on stock markets* (SAFE Working Paper 310).

Naughton, J. P., Wang, C., & Yeung, I. (2019). Investor sentiment for corporate social performance. *The Accounting Review, 94*(4), 401–420.

Pizzutilo, F., Mariani, M., Caragnano, A., & Zito, M. (2020). Dealing with carbon risk and the cost of debt: Evidence from the European market. *International Journal of Financial Studies, 8*(4), 61.

Sassen, R., Hinze, A. K., & Hardeck, I. J. (2016). Impact of ESG factors on firm risk in Europe. *Journal of Business Economics, 86*(8), 867–904.

Schettini, G. J. (2005). *An overall view of the rating activities on Italy carried out by AEI standard ethics*, Notizie di Politeia. Rivista di Etica e Scelte Pubbliche, 2005, XXI, 78. Le Prospettive Della Responsabilitá Sociale In Europa. Studi, Ricerche e Casi, Emilio D'Orazio (edited by), (pp. 255–274).

Schettini, G. J., & Schettini, G. L. (Eds.). (2002). *Etica, futuro e finanza*. Il Sole 24Ore, Milano.

Scholtens, B., & Witteveen, E. (2021). Shocks, stocks and ratings: The financial community response to global environmental and health controversies. *Global Environmental Change, 68*, 102245.

Sen, A. K. (1991, April). *Money and Value: on the ethics and economics of finance*. Paolo Baffi lectures on Money and Finance, Banca d'Italia, Rome.

Serafeim, G. (2018). Public sentiment and the price of corporate sustainability. *Financial Analysts Journal, 76*(2), 26–46.

Serafeim, G., & Yoon. A. (2021). *Stock price reactions to ESG News: The role of ESG ratings and disagreement* (Harvard Business School Accounting & Management Unit Working Paper 21-079).

Starks, L. T., Venkat, P., & Zhu, Q. (2017). *Corporate ESG profiles and investor horizons*. https://doi.org/10.2139/ssrn.3049943

CHAPTER 6

ESG Factors in the Investment Process

Giovanna Frati, Julien Mazzacurati, and Barbara Alemanni

Introduction

Sustainable investing can be broadly defined as incorporating environmental, social and governance (ESG) factors into investment decision-making. Rather than displacing conventional financial investing, ESG investing complements it by introducing non-financial dimensions: ethical

Barbara Alemanni authored Box 6.1.

G. Frati (✉)
Commissione Nazionale per le Società e la Borsa (CONSOB), Milan, Italy
e-mail: g.frati@consob.it

J. Mazzacurati
European Securities and Markets Authority (ESMA), Paris, France
e-mail: julien.mazzacurati@esma.europa.eu

B. Alemanni
Bocconi University, Milan, Italy
e-mail: barbara.alemanni@unibocconi.it

University of Genoa, Genoa, Italy

N. Linciano et al. (eds.), *Information as a Driver of Sustainable Finance*, Palgrave Studies in Impact Finance,
https://doi.org/10.1007/978-3-030-93768-3_6

and social value in addition to financial value; criteria for screening, tilting or benchmarking existing portfolios; and the integration of ESG metrics which can be combined with traditional measures of financial performance.

ESG investing can take many different forms even as the market appears to have coalesced over the years around a number of implementation strategies. This chapter starts by presenting the main sustainable investment strategies, before providing an overview of recent market developments.

As already said, there has been a paradigm shift in recent years, with large swathes of global financial markets turning to sustainable investing, particularly in Europe, and so-called "green finance". There are several underlying drivers explaining this trend, among which growing awareness of the climate urgency, greater attention paid to social issues (especially from Millennials), and a long-term shift from shareholder to stakeholder capitalism. The COVID-19 crisis further served to accelerate the transition to sustainable investing.

The financial sector in general, and the asset management industry in particular, has responded in kind to booming demand for sustainable investment products by their customers, including retail and institutional investors, seeing in this its own long-term interest—if not survival. Concretely, this has materialized in the form of a very large number and products and vehicles with sustainability characteristics or objectives being launched, such as investment funds with an ESG mandate.

In order to ensure consistency between the stated investment policy and actual investment choices, regulatory safeguards relating first and foremost to the investment process are indispensable.

In the EU regulatory framework, the institutional investors (or "asset owners" or "products") are portfolios managed by an authorized[1] and supervised entity (the "asset manager") and subscribed by the final investors. The asset owner can assume the form of either a collective investment scheme or a portfolio managed on discretionary basis or an insurance product or pension funds.[2]

The investment policy and the investment strategy of the asset owner are disclosed in the undertaking's regulation (or contract) and in the offering documents[3] and turn to be the mandate attributed by the final investors to the asset manager. Accordingly, the asset manager shall put in place investment choices on the basis of that mandate.

The strategic objective (or strategic behaviour) of the asset manager should be the provision of products matching the final investors' needs, being able to ensure simultaneously profit for both asset owner and asset manager.

The achievement of strategic objective requires the asset manager to adopt a dedicated organizational structure (the so-called, operational behaviour) allowing to perform a fair and effective investment process.

Indeed, the decision-making process or, briefly, investment process is the core business of the asset manager and could be regarded as the key allowing the manager to convert the product investment policy as disclosed in the offering documents (the mandate) in concrete investment choices.

Obviously, both organizational structure and investment process adopted by the asset manager should be designed in light of the peculiarities of the managed products. Hence, the strategic decision of providing investors with the ESG products should be properly mirrored in the operational structure in order to allow the implementation of investment choices consistent with ESG factors.

Despite these requirements, the breath-taking pace at which sustainable investment products are being launched and large amounts of money being allocated to these, combined with the absence of common definitions and transparency requirements until recently, have fuelled concern over the actual sustainability credentials of these products and their broader impact on ESG matters. More specifically, there are concerns that fund managers' communication is not proportionate to the sustainability credentials of their ESG products, nor truly representative of the impact of their investments.

This problem is compounded by broader issues regarding the quality and reliability of ESG data. To this day, the disclosure of non-financial information remains sparse, concentrated in mature markets, and usually limited to larger firms. To fill these gaps, fund managers rely on proxies or alternative information from third-party data firm, such as ESG rating providers (see Chapter 5). However, these data come with their own set of issues, including their lack of comparability and limited transparency, raising further questions as to the reliability of the sustainability assessment of investment choices being carried out on their basis (see also Chapter 3).

In light of the mentioned background, below the following items are analysed: the evolution of the ESG factors (section "ESG Factors and Sustainable Investing"), the main drivers/requirements—from a regulatory point of view—of investment process of the manager of ESG products (section "The ESG Manager and the Decision-Making Process"); the distinctive characteristics—from a regulatory point of view—of the ESG products (section "The Asset Owner: ESG Products and the Role of the Benchmark"). The last sub-section is dedicated to the representation of the main challenges for the asset management deriving from ESG factors (Section "The Main Challenges").

ESG Factors and Sustainable Investing

The term sustainable finance covers a large array of ESG factors. As discussed in Chapter 2, while there is no agreed definition specifying exactly what an ESG factor is, nor which types of issues may or may not be considered, there is nonetheless implicit agreement over the categorization of certain issues in each of the environmental, social and governance pillars (CFA Institute, 2015):

- The environmental pillar encompasses climate-related issues, as well as pollution, biodiversity, energy, and waste and natural resource management.
- The social pillar covers issues related to gender and ethnic diversity, economic development, inequalities, healthcare, community and labour relations, and human rights.
- The governance pillar captures both institutional governance and corporate governance issues, spanning issues such as corruption, lobbying and, as well as board composition, company ownership structure, shareholder rights and executive compensation.

In practice some of these themes overlap, which can lead asset managers and data providers to categorize the same issues under different pillars. This is the case in particular for themes that stride social and governance aspects such as employee compensation or labour relations. Moreover, there are clear interlinkages between other themes, e.g. climate change and inequalities, implying a degree of correlation (if not causality) between the different pillars—rendering any isolated assessment of one of the three pillars a necessarily partial and incomplete exercise.

One element that these factors have in common is that they have a *material* impact on the financial performance and/or risk profile of companies. Related to this, a fundamental aspect of ESG investing is the so-called financial materiality assessment, which is intended to quantify the materiality or impact of ESG factors on firms. This materiality assessment can take many different forms, and have a significant impact. To promote a common approach, a set of industry standards and best practices (e.g. metrics to be used) for corporate financial materiality assessment have been developed (Sustainability Accounting Standards Board—SASB, 2018; see Chapter 4 for further discussion on this).

Sustainable Investing: The Context

As highlighted in Chapter 2, there has been a paradigm shift in recent years in favour of sustainable investing, particularly in Europe. Originally a niche in the asset management industry known as socially responsible investing (or SRI),[4] ESG investing has now become pervasive, concerning all financial sectors, and a growing number of financial instruments, products and vehicles. Although estimates vary greatly depending on the definition used (see below), the total value of assets following sustainable strategies in 2020 was estimated to be as high as €45 trillion (J.P. Morgan, 2020).

Box 6.1: by B. Alemanni ESG versus SRI
Under the umbrella of sustainable investing there are very different ideas of investment approach. As a blanket investment term, sustainability has become a catch-all for a company's efforts to "do better" or "do good". This investment approach is best defined by the three pillars of sustainability: economic growth, environmental protection, and social progress, also referred to as "people, planet, and profits". In a nutshell, sustainable investing directs capital to companies fighting climate risk and environmental destruction, while promoting corporate responsibility. In this

general framework it is important to specify the difference between SRI and ESG investing, particularly as naming conventions are somewhat confused in this area (Eccles & Viviers, 2011).

SRI has many definitions but is broadly considered to be an investment style that combines financial goals with social and environmental objectives (Sparkes, 2001). Historical records claim that such investments started with the Religious Society of Friends (Quakers) in 1758 when the Quaker Philadelphia Yearly Meeting prohibited members from participating in buying or selling humans as part of the slave trade. One of the early adopters was Methodist founder John Wesley, whose sermon "The Use of Money" outlined his basic tenets of social investing: not to harm your neighbour through your business practices and to avoid industries like tanning and chemical production, which can harm the health of workers.

In general, SRI investors encourage corporate practices that are morally grounded and promote environmental stewardship, consumer protection, human rights and racial or gender diversity. For example, some socially responsible investors avoid investing in businesses perceived to have negative social effects such as alcohol, tobacco, gambling, pornography, weapons and fossil fuel production. Essentially, for socially responsible investors morality is the bottom line.

SRI place a premium on positive social change and weigh both financial returns and moral values in investment decisions. This method of investing is based on selecting socially responsible companies and building portfolios and excluding companies negatively impacting society or the environment or that have a morally corrupt value system.

In contrast, ESG focuses on three specific, foundational pillars that are crucial to today's corporate management and investors alike. Environmental issues can include pollution, climate risk, exposure to extreme weather, carbon management and use of scarce resources. Social issues can include product safety, human rights, worker safety, customer data protection and diversity and inclusion. Governance issues can include factors such as accounting standards compliance, succession planning, anti-competitive behaviour and a strong ESG management process.

ESG data and metrics are used to gain insights into the success and value of a company's performance and policies in order to mitigate risk and identify superior risk-adjusted returns. Essentially, the focus of ESG investing is on increasing the bottom line through investments in responsible companies that are being well managed.

ESG investing has a singularly financial motivation. ESG investors base decisions on a broader set of criteria that are not influenced by moral decision-making or exclusive to environmentally and socially conscious business practices. Rather than completely rule out companies and industry sectors based on blanket criteria such as alcohol, animal testing, or weapons, ESG investors identify and rank businesses based on data and consideration of how ESG risks and opportunities can impact the company's performance. Thus ESG investing avoids the legal conundrum of whether the consideration of social or environmental factors is consistent with beneficiaries' best interests, as its core focus is financial, and therefore economically rational (Gray, 2012). The term also avoids the values-based judgements necessarily present in SRI, as well as the implication that investments that are avoided are "irresponsible" (Gray, 2012).

The underlying drivers of this trend are multiple and diverse. As highlighted also in the previous chapters, some structural developments can help understand the root causes for the massive reallocation of capital to sustainable investment products that has taken place in recent years. First and foremost, the 2015 Paris agreement has crystallized minds around the urgency of the fight against climate change, and the responsibility that humanity bears in this respect, as well as the role it needs to play to mitigate its effects. Second, the consequences of the 2007–2008 global financial crisis on employment, inequalities and more broadly, the social fabric, have led to growing attention being paid to social and environmental issues, further accentuated by generational changes (Pew Research Center, 2021). Last but not least, a long-coming shift away from shareholder capitalism and towards stakeholder capitalism is leading a growing number of businesses to see the long-term value in attending to the need of their employees, customers and business partners, rather than focusing exclusively on delivering short-term value to shareholders. Taken together, these developments—combined with growing media attention—help understand why ESG factors appear to have taken centre stage in just a few years, including in financial investing.

The short-term drivers of demand for sustainable investment products are equally varied. As discussed in Chapter 7, some investors simply aim to improve social or environmental outcomes through their investments. Others focus on identifying ESG issues to reduce portfolio risks or to identify undervalued opportunities. Perhaps more importantly, many businesses and investors also see the potential reputational risk that may come either from being associated with unethical practices, or simply being perceived as not being sufficiently active in this space. Together with the need to cater to their clients, the latter point also explains asset managers' enthusiastic embrace of ethical investing principles.

Sustainable Investing Strategies

There is a wide array of strategies that can be described as sustainable, reflecting conflictual views as to the very definition of what sustainability means. The absence of clear, common definitions leads in turn to multiple challenges: measuring the market share and growth of different strategy types; assessing the impact that these strategies have on environmental and social outcomes; and understanding their micro implications for risk management purposes, as well as their broader effects on financial stability.

Despite some remaining differences, the market appears to have coalesced around a few broad themes which are illustrated in Table 6.1, based on a bi-annual survey from the European Sustainable Investment Forum (Eurosif).[5]

Sustainable investing strategies can be purely qualitative or quantitative, or combine elements of both. Strategies based on exclusions are mainly qualitative: these exclude companies or sectors from the investment universe due to their reputation (based for example on existing controversies) or the nature of their activities (e.g. tobacco, weapons, etc.). Similarly, norms-based screening—which can be either positive or negative—relies on a binary assessment of companies' compliance with

Table 6.1 Sustainable investment strategies

Strategy	*Description*
Best in class	Picks companies with best ESG scores
Thematic	Selects assets around a particular theme
Norms-based screening	Selects assets based on compliance with global standards and norms
ESG integration	Integrates ESG factors in investment analysis
ESG engagement	Favours ESG stewardship and investor activism
Exclusions	Excludes assets based on specific criteria
Impact investing	Focuses on social or environmental impact

Source Eurosif (2018)

global standards (e.g. the UN Principles for Responsible Investing) or alignment with international agreements (e.g. the Paris agreement).

Qualitative strategies have been historically dominant in the European fund industry, with a combined €15 trillion in assets under management in 2015. However, they have receded in recent years as investors turned in growing numbers to ESG-based strategies, also perceived to have a greater impact. For example, ESG engagement is another form of qualitative strategy, whereby asset managers engage directly with the investee companies in order to drive internal changes—which is increasingly perceived as a more efficient means than simple divestment to help bring about improvement in various ESG areas.

Quantitative strategies have gained popularity in recent years, in line with the growing role of data and analytics in the investing business, also thanks to the availability of new technologies and growing computing power. Best-in-class and worst-in-class strategies, for example, rely on ESG scores to identify companies within sectors and/or countries with the best or worst ESG credentials. One strategy in particular (ESG integration) has experienced widespread adoption in the asset management industry, reflecting the diversity of practices captured as well as its flexibility.[6] ESG integration can take place at different stages of the investment process, including valuation (e.g. identifying potential mispricing or arbitrage opportunities from an ESG-based assessment) or portfolio construction (e.g. calculating weights based on the financial materiality assessment of ESG factors).

Finally, despite their relatively modest size relative to the broader sustainable investment universe, impact investing (which aims to achieve quantitative targets) and thematic investing (focusing on particular sectors) strategies have enjoyed a very strong momentum of late, reflecting growing doubts as to the environmental and social impact of strategies based on ESG ratings, and of more broadly the reliability of ESG factor-based assessments see Section "The Main Challenges".

ESG Fund Market Trends

In an illustration of the remarkable growth of ESG investing strategies in recent years, the total value of assets that are managed globally by funds with an ESG mandate (ESG funds, henceforth) has grown by 170% between 2015 and 2020, with more than half of these domiciled in the Euro Area. In the European Union as a whole, the total assets under managements of ESG funds amounted to around €1500 billion in mid-2021, or 18% of the industry's total assets in the region (see Fig. 6.1). This trend has accelerated over time, with Q1 2020 marking a turning point as the COVID-19 crisis further brought attention to sustainability-related issues. This shift has taken place to some extent at the expense of traditional (non-ESG) funds: in the first six months of 2020 only, European ESG equity funds experienced inflows of €14 billion, compared with *outflows* of €29 billion from non-ESG peers.

Until recently, the market for ESG funds was dominated by actively managed funds (also known as stock-picking; more on the definition of active and passive funds in section "Recent Developments in ESG Markets" in Chapter 2). This reflected multiple factors, including the flexibility to invest in companies seen as performing relatively better on ESG factors, more engagement opportunities with the investee company, but also the limited number of ESG benchmarks available in the market. However, the trend appears to have reversed in the course of 2020, with

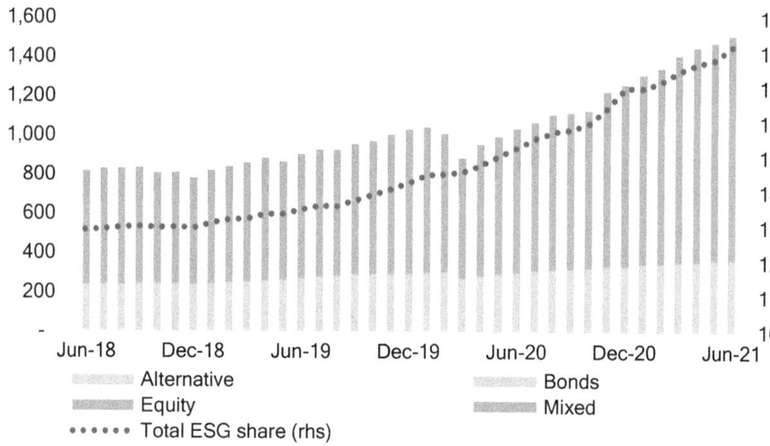

Fig. 6.1 Graph: Assets under management of EU ESG funds (*Note* Assets under Management [AuM] of EU-domiciled ESG funds by type of fund, EUR billion, and share of ESG fund AuM in total industry AuM [right axis], in %. *Source* Morningstar, Refinitiv Lipper, ESMA)

a growing number of passive investment products (including exchange-traded funds or ETFs) becoming available in the market and their assets under management trebling in 2020 in Europe.

Aside from the management type, another recent albeit noteworthy development of the ESG fund industry is the diversification of assets managed by ESG funds. The ESG space remains essentially composed of equity funds, reflecting greater underlying liquidity and information availability, but ESG corporate bond funds—including ETFs—and alternative investment products have made some in-roads in 2020. While the available evidence still points to a relatively small share of the global hedge fund market (around 3% according to BarclayHedge), recent surveys show that a growing number of managers are either already integrating or planning to integrate them into their strategy (Barclays, 2021).

The ESG Manager and the Decision-Making Process

The Decision-Making Process in the UCITS Directive and the AIFMD

Asymmetries in information are inherent in the relationship between the asset manager and the final investors subscribing the financial products (as said before, the product is the "asset owner"). Therefore, the EU legislator has regulated the provision of portfolio's management and the decision-making process with the objective to prevent potential misconduct from the asset manager side and to mitigate the presence of conflict of interests.

Specifically, the decision-making process turns to be regulated in the European legal framework dedicated to the collective investment activity—UCITS Directive and AIFMD[7] and their delegated acts (UCITS Delegated Directive and AIFM Regulation)[8]—taking into account that the entities authorized under UCITS Directive and/or AIFMD are properly dedicated to the provision of management activity.

On the basis of the mentioned framework, the decision-making process—bearing in mind that it, taken as a whole, shall be inspired by the best interest of the final investors—is broken down in a number of steps (Table 6.2).

Specifically:

Table 6.2 Steps included in the decision-making process

Steps	*Description*
Definition of investment policy and strategy	The investment policy and strategy represent the perimeter for the asset allocation and they are defined in the offering documents
Due diligence	The acquisition of adequate knowledge and understanding of investible assets is a necessary step in view of the investment choices implementation
Investment decision	After the due diligence, the investment choices consistent with the investment policy are put in place
Monitoring	The portfolio composition shall be checked on periodic basis in view of ensuring the proper implementation of the investment policy and strategy

1. the definition of the investment policy and strategy for each managed fund.[9]

The investment policy and strategy—disclosed in the regulatory documents—provide the asset manager with the perimeter (in terms, inter alia, of asset classes, geographical or sectoral exposure, credit, market and liquidity risk, market risk) within which it can take the investment choices;

2. the due diligence and the consequent investments decisions.[10]

The due diligence entails the acquisition of adequate knowledge and understanding of the assets investible by the fund in order to put in place investment choices in compliance with the objectives, the investment strategy and the risk limits of the fund; the due diligence is followed by the implementation of concrete investment choices;

3. the monitoring of implemented investment choices.[11]

The monitoring, from a risk management perspective, shall be carried out on a periodic basis, with the aim of checking that the investment policy and strategy and the risk limits of each managed funds are properly and effectively implemented and complied. To this end, the asset pricing is the crucial variable to measure the impact of each invested financial instrument on the risk-return of the portfolio.

Due to the asymmetric information, the conflicts of interests potentially affect each step of decision-making process. Instead of banning them, the regulator has accordingly established a dedicated framework aiming at their management and mitigation. On the basis of regulatory approach, situations of conflict of interests shall be, firstly, identified by the manager and, afterwards, managed through the adoption of appropriate organizational measures (such as, for instance, Chinese walls).[12]

Following the 2018 Action Plan on sustainable finance, the EU legislator acknowledged the relevance of both ESG factors in the decision-making process of the asset managers by amending the delegated Acts of UCITS Directive and AIFMD, as it will be discussed in Chapter 8.

The Engagement in the Investee Companies

While some funds cannot increase their exposure to the invested companies beyond a pre-defined limit (e.g. on the basis of art. 56 of UCITS Directive, an investment company or a management company acting in connection with all of the common funds which it manages and which fall within the scope of the UCITS Directive shall not acquire any shares carrying voting rights which would enable it to exercise significant influence over the management of an issuing body), the investment objective of the other investment funds that are not subject to such a limit (e.g. private equity funds) is properly identified in the acquisition of the governance of the target companies.

However, regardless of the size and the relevance of the exposure in the investee companies, the monitoring of achieved investments makes up a key part of the investment process and, about that, the Directive (EU) 2017/828 (henceforth, Shareholder Rights Directive II or SHRD II) points out the need to supervise the investee companies "on the relevant matters", including, inter alia, "*social and environmental impact and corporate governance*".[13]

The SHRD II perspective of the monitoring of the implemented investment choices is something new compared to the monitoring envisaged by the Directives dedicated to the collective management activity (UCITS and AIFMD). Indeed, the latter focus mainly on the need of monitoring the implemented investments to ensure, on ongoing basis, their consistency with the investment policy and strategy described in the offering documents; while in the Shareholder Rights Directive II the monitoring is represented as a crucial step to enhance the governance of the companies and their ESG impact.

More in detail, the Shareholder Rights Directive II Directive recognizes that the role performed by the institutional investors and asset managers in the corporate governance of investee companies is of the utmost importance in view of promoting a long-term performance and comes up with the conclusion that "*Effective and sustainable shareholder engagement is one of the cornerstones of the corporate governance model of listed companies, which depends on checks and balances between the different organs and different stakeholders. Greater involvement of shareholders in corporate governance is one of the levers that can help improve the financial and non-financial performance of companies, including as regards environmental, social and governance factors, in particular as referred to in the Principles for Responsible Investment, supported by the United Nations* [...]".[14]

In other words, pursuant to Shareholder Rights Directive II, both short-term approach and lack of engagement with the investee companies could determine a suboptimal level of investments (for example in research and development) and the detriment of the long-term performance of both the companies and the investors.

Accordingly, the regulatory approach adopted by EU legislator in the Shareholder Rights Directive II (and even in Disclosure Regulation) bolsters the position pursuant to which fiduciary duty, far from being a barrier, represents a positive input to integrate ESG factors in the investment process (Freshfields Bruckhaus Deringer, 2005; PRI et al., 2015). Pursuant to the mentioned approach, failure in the consideration—in the context of investment choices—of long-term investment value drivers (including ESG factors) turns out to be a failure of investment duty. The result should be a portfolio composition significantly allocated towards well-governed companies in view of contributing to the goals of a greener economy and a more sustainable society.

In the abovementioned background, the tools identified by the EU legislator in order to stir a long-term approach and, through it, to enhance the implemented investments are mainly based on the adoption of the engagement policy and on the disclosure to investors.

The engagement policy, hinging on long-term strategies with the investee companies, should aim at aligning the interests between the final beneficiaries of institutional investors, the asset managers and the investee companies.[15]

The disclosure to investors should aim at achieving a good standard of investors' awareness, enabling ultimate beneficiaries (such as future pensioners) to optimize investment decisions, facilitating the dialogue between companies and their shareholders, encouraging shareholder engagement and strengthening their accountability to stakeholders and to civil society.[16]

The Asset Owner: ESG Products and the Role of the Benchmark

Management Approaches (Active vs Passive) and Investment Policy

The Directives on collective investment schemes (UCITS directive and AIFMD), being focused on entities whose objective is the asset management activity, provide also with a reference to the viable management approaches and investment policies.

Indeed:

- as far as the management approaches are concerned, the Level 3 measures[17] of the UCITS Directive[18] single out: market funds, absolute return funds, total return funds, life cycle funds, structured funds[19];
- as far as the investment policy are concerned, the Level 2 measures of AIFMD[20] refer to: hedge funds, private equity funds, real estate funds, fund of funds.

Even if the abovementioned management approaches and investment policies are identified by the regulator in reference to, respectively, UCITS funds (i.e. funds compliant with the risk spreading rules set out in the UCITS Directive) and AIFs (i.e. alternative investment funds, that shall be intended as collective investment schemes different from UCITS funds), they can be helpfully considered—broadly speaking—for all kind of institutional investors.

What stems from the classification of management approaches highlighted in the UCITS framework is the substantial discrimination between funds managed with respect to a benchmark (so-called market funds), on one side, and funds that are not (so-called flexible funds), on the other side.

ESMA has recently clarified the meaning of a management approach including or implying a reference to a benchmark.[21] Specifically, ESMA introduces a list of indicators potentially signalling the benchmark index plays a role in the management of the UCITS, for example, in the explicit or implicit definition of the portfolio's composition or in the UCITS performance objectives or in the UCITS performance measurement.

These indicators include, inter alia, the following situations: the fund uses a benchmark index as a universe from which to select securities; the fund portfolio holdings are based upon the holdings of the benchmark index; performance fees are calculated on the basis of performance against a reference benchmark index.[22]

In any case, when a fund is managed with respect to a benchmark, the degree of freedom from the benchmark gets the relevant driver to identify the management style: active style vs passive style.

The passive management style (or index tracking) is achieved when the fund is substantially aligned with the benchmark in terms of risk-reward profile, asset allocation and portfolio composition.

The passive management style could be put in place through a physical replication or a synthetic replication.

Indeed, the traditional way to replicate an index is to hold directly all the underlying index constituents (physical replication). By using such a structure, the investment manager has to rebalance the portfolio in line with periodic changes as stipulated by the index provider.

Alternatively, funds can use swaps to replicate the performance of the index. In order to provide the investors with a return linked to the performance of the underlying index, the fund invests in transferable securities (Invested Assets) and uses derivative techniques such as swap agreements, to exchange the performance of the Invested Assets against the performance of the index.

On the opposite side, a fund is intended to be actively managed when a certain degree of freedom from the benchmark (from the perspective of asset allocation, stock-picking, risk-reward profile) is spotted.

The Peculiarities of ESG Products

Bearing in mind that the discrimination between market funds and flexible funds shall be considered valid also for ESG funds, the peculiarity of such funds (and, broadly speaking, the peculiarity of the ESG products) is identifiable in the investment objective.

Indeed, the objective of ESG products is made up of sustainable investments and/or investments contributing to the environmental or social characteristics.

Anticipating the review of the implementation of the 2018 Action Plan discussed in Chapter 8, it is useful to recall that the Sustainable Finance Disclosure Regulation or SFDR, art. 2(17), defines the sustainable investments ("*an investment in an economic activity that contributes to an environmental objective [...] or [...] to a social objective [...] provided that such investments do not significantly harm any of those objectives and that the investee companies follow good governance practices, in particular with respect to sound management structures, employee relations, remuneration of staff and tax compliance*") and envisages two categories of ESG products:

- the first category refers to the products promoting environmental or social characteristics[23];
- the second category concerns products whose objective is the sustainable investments.[24]

Disclosure Regulation, in identifying such categories, does not provide with a definition; however, the delegated act of Disclosure Regulation has clarified that the two categories of ESG products can be discriminated on the basis of the investible assets and investment strategies.

Indeed, while the investible assets of funds with sustainable objective should be limited to the sustainable investments, funds with environmental or social characteristics can be invested in a wider range of assets, including assets qualifiable as sustainable investments, assets contributing to specific environmental or social characteristics and, even, assets that are not relevant to the environmental or social characteristics promoted by the fund.[25]

Likewise, also the range of investment strategies achievable by funds with sustainable characteristics turns to be wider than that achievable by funds with sustainable objective: funds with sustainable objective are specifically designed to target the sustainable investments while funds promoting environmental and/or social characteristics cover various investment approaches and strategies (from best-in-class to specific sectoral exclusions).

Having said that, it is clear that the risk of greenwashing gets more relevant for funds with sustainable characteristics rather than for funds with sustainable objective in consideration of the broader covered set of investible asset classes and investible strategies.

The Benchmark Regulation and Its Amendments to Include Climate-Related Benchmarks

When ESG products (in both form of products with sustainable characteristics and products with sustainable objective) are managed in reference to a benchmark,[26] the fund's portfolio composition, its risk/return profile and, where relevant, the computation of performance fees are influenced by the benchmark.

However, benchmarks are subject to the risk of manipulation, since conflicts of interests are inherent in the benchmark provision process. Broadly speaking, any discretion that can be exercised in providing input data and in benchmark computation creates an opportunity to manipulate a benchmark.[27]

Potential benchmarks failures can undermine investors' protection and market integrity. That's why the European regulator, with the view to strengthen the benchmark's reliability, in 2016, has intervened with an ad hoc regulation (the Benchmark Regulation[28]).

The main measures identified by the regulator to reach that goal are:

- the authorization or registration and supervision of benchmark administrators, since the administrators control the benchmark determination process.

In order to be authorized or registered, the administrators are required to implement adequate governance arrangements to control conflicts of interest and to safeguard confidence in the integrity of benchmarks[29];

- specific requirements concerning the input data and the methodology on the basis of which the benchmark is computed.

Inter alia: the input data shall be sufficient to represent accurately and reliably the market or economic reality that the benchmark is intended to measure and where a benchmark is based on input data from contributors, the administrator shall obtain, where appropriate, the input data from a reliable and representative panel or sample of contributors so as to ensure that the resulting benchmark is reliable and representative of the market or economic reality. The methodology shall be robust, traceable and verifiable and its key elements shall be disclosed;

- Ad hoc requirements for specific categories of benchmark, critical benchmark,[30] whose cessation could cause losses to investors and impact financial stability.

About that benchmarks, in order to preserve their existence, the legislator has recognized a power for the relevant competent authority to require mandatory administration, as well as to require mandatory contributions.

In 2019, the benchmark Regulation has been amended[31] to take in consideration the developments of financial products referencing to benchmark focused on environmental objective: the definitions of EU Climate Transition Benchmark[32] and EU Paris-aligned Benchmark[33] have been included in the Regulation with the objective to facilitate investments in sustainable projects and assets across the EU (see also Chapters 8 and 10).

In the mentioned framework:

- the SFDR has clarified that products with sustainable investment objective that rely on an index to attain the sustainable investment objective are passive products and they must demonstrate that the designated index is aligned with the product's sustainable investment objective (art. 9(1));
- the level 2 measures of Disclosure Regulation has envisaged that "*Financial products should not pursue low-carbon investment objectives without using new Union climate-related benchmarks. If such benchmarks are not available, financial market participants should demonstrate how the financial product complies with the relevant standards applicable to EU Paris-aligned benchmarks or EU Climate*

Transition benchmarks as set out in the Regulation (EU) 2019/2089" (recital 29).

The Main Challenges

ESG Funds

The entry into force of SFDR requirements in March 2021 is an essential milestone in global efforts towards increased transparency of products with sustainability characteristics or objectives in general, and funds with an ESG mandate in particular. However, it will require some time for market practices to converge towards unified definitions and coordination among national authorities to ensure a consistent and harmonized approach to ESG fund supervision. Until such clarifications have been made, the universe of ESG funds can be broadly described as consisting of two types of funds: funds with an ESG label, and non-labelled funds with an ESG strategy described in their documentation.

ESG labels play a useful role in offering a degree of guarantee that minimum sustainability criteria are met by the fund. In turn, fund managers see in ESG labels a potent means to signal to investors the credibility of their commitment to ESG matters. However, a high degree of heterogeneity exists in ESG fund labels across EU countries, reflecting the absence of an EU-wide ESG label. ESG fund labels exist for example in Austria, France, Germany and Luxembourg but rely on different criteria. In some EU countries, ESG labels do not exist. These country-specific situations raise consistency issues when funds are marketed beyond national borders. Recognizing this problem and the potential for creating a strong EU-wide brand for green funds, the European Union is working on the development of an *Ecolabel* for retail investment products (European Commission, Joint Research Center, 2021). Under the current technical proposals, retail investment products such as UCITS and retail AIFs could qualify for the Ecolabel provided that they meet the following five criteria[34]:

1. **Investing in environmentally sustainable economic activities**: A minimum proportion of assets under management should be invested in economic activities making a substantial contribution to environmental objectives as defined under the EU Taxonomy Regulation.

2. **Exclusions based on environmental aspects**: Investment funds are barred from holding equities or bonds issued by companies deriving more than 5% of their revenue from environmentally harmful activities, or from holding sovereign bonds issued by countries that are not party to the Paris agreement and other international conventions on the environment.
3. **Exclusions based on social and governance aspects**: Investment funds are also barred from holding equity or bonds issued by companies that do not comply with minimum conditions (in line with global standards) or derive revenues from certain socially harmful activities, or sovereign bonds issued by countries that have not ratified international conventions on labour rights and corruption.
4. **Engagement**: Fund managers need to have a documented engagement policy in line with the EU Taxonomy environmental objectives, exercise voting rights to fulfil the objectives of their engagement policy and regularly engage at management level with companies within their portfolio.
5. **Measures taken to enhance investor impact**: Fund managers must report on the impact of their product to inform investors of the measures they have implemented.

In the absence of label, funds can alternatively describe in their documentation (prospectus, KIID, etc.) the sustainability component(s) of their strategy. However, the absence of rules before SFDR implies that there have been various problems from this practice. First, the wording used by fund managers can vary greatly from one to the other, both in terms of length and vocabulary, rendering any comparison virtually impossible. Second, unverifiable claims on the environmental or social impact of funds can be made without any guarantee that the fund achieves its objectives (nor that it has any impact). Third, this approach does not allow for the benchmarking of sustainability performance against objective criteria since none have been set.

In the absence of official EU-wide definitions or label (notwithstanding the development of a European Ecolabel for retail investment products, including so-called "green" investment funds (see Chapter 4) a fundamental question has been the extent to which funds' ESG communication could influence the perception that investors have of these funds. Aside from national labels, rules have been in place for some time in several EU countries on the way and extent to which funds are allowed to communicate their ESG strategies,[35] but these rules are neither systematic nor harmonized.

Of the nearly 2300 new equity, bond and mixed funds launched in 2019 and 2020 in Europe, almost a third made reference in their regulatory or marketing documentation to ESG investing (as defined in section "EU Regulation and Retail Investors' Sustainable Investments" in Chapter 7).[36] In the twelve months following their inception dates, ESG equity funds attracted on average EUR 95 mn, compared with EUR 70 mn for non-ESG funds.[37] Rather than launching new funds, some asset managers are opting instead to onboard ESG strategies within existing funds and revise the documentation of their funds accordingly to incorporate ESG elements—thereby making it possible to identify their conversion dates. This conversion effect seems to pay off: based on a sample of 880 equity and mixed funds, equity funds attracted on average five times more net flows in the 10 months after conversion than in the 10 months before, while mixed funds were able to stem outflows.

These observations, combined with broader evidence on recent trends in ESG investing, highlight the dire need for rules in this area. Transparency requirement should help investors to compare and assess the products they are being offered. The establishment of clear and measurable criteria, for example in the form of an EU-wide label, would not only serve to identify funds with the highest impacts, but also serve to establish a benchmark against which the performance of non-labelled funds could be measured.

ESG Ratings and ESG Funds

To address these issues, several data providers have introduced portfolio-level ESG ratings, which are based on the ESG rating of individual portfolio securities held by the fund (see also Chapter 5). These fund ESG ratings can be useful to investors, in that they allow for greater comparability and transparency of the sustainability credentials of funds' portfolio holdings based on an objective third-party assessment. However, as extensively discussed in Chapter 5, fund ESG ratings also have their own limitations: they rely on purely quantitative criteria but ignore qualitative aspects such as fund manager's engagement with shareholders or companies; they create possible biases against non-rated companies; and they rely on ESG ratings, which are widely viewed as inconsistent across rating providers, if not unreliable.

Nonetheless, such fund ESG ratings are a clear progress from the complete absence of information. There also appears to be a degree of overlap between funds claiming to pursue sustainable strategies and funds with high ESG ratings: out of the roughly 2000 ESG funds in the EU that had an ESG rating in 2020, around 65% were rated "High" or "Above average". This compares with 37% for non-ESG funds with an ESG rating—also highlighting that a relatively large number of funds receive a favourable ESG rating for unrelated reasons (e.g. sectorial allocation, benchmark tracked, geographical focus, etc.).

Questions that relate specifically to the impact of ESG funds closely intertwine with an ongoing debate about the reliability of ESG ratings. ESG ratings are useful in that they allow investors to obtain an external opinion on the sustainability credentials of companies and countries, but also investment fund portfolios. ESG rating providers can play a valuable role through structured assessments based on rigorous methodologies, extracting value across thousands of issuers from ESG disclosures through their rating process (Boffo & Patalano, 2020). Thus, they have an important role to play in facilitating the transition to a more sustainable economy, allowing the optimal allocation of capital to cleaner companies and infrastructures. In addition, ESG ratings and providers incentivize companies to improve their ESG credentials and levels of disclosure.

However, these benefits are hampered by several concerns. The absence of a common definition for ESG ratings leads to investor and issuer confusion and misunderstandings. While the existence of different methodologies and approaches reflects the diversity of client needs, varying degrees of transparency further limit the comparability of these ratings and the ability to understand what the main drivers and limitations are. This divergence between ESG ratings is documented for example by Berg et al. (2019), who use ratings from five prominent ESG rating providers. They find that ratings are only 60% correlated, compared with 99% for credit ratings from the largest credit rating agencies, which is mainly explained by differences in measurement (i.e. measuring the same object in different ways) and aggregation rules. A high level of disagreement between ESG ratings is confirmed by Billio et al. (2020), who attribute this problem to a lack of commonality in the definition of environmental, social and governance components. Disagreement leads in particular to discrepancies among ESG indices, with very low agreement rates on the constituents of comparable ESG indices (in terms of coverage and sector composition) from four different providers, even after controlling for geographical differences.

The reliability and comparability concern around ESG ratings matters enormously when considering the potential impact of ESG rating-based products, including many ESG funds, on environmental and social aspects. For example, Boffo et al. (2020) find a positive correlation between high environmental scores from one provider and high level of CO_2 emissions and waste (i.e. some highly rated firms are associated with higher levels of CO_2 emissions). In the same vein, a recent study highlighted that a third of the 33 climate funds sold in the UK are invested in oil and gas companies. Other examples of ESG rating divergence are not lacking, with Tesla frequently cited as receiving a top ESG score from one ESG rating provider while another gives it the lowest grade.

This lack of consistency is particularly problematic for ESG indices, tracked by a large number of passively managed funds and ETFs, with analysis showing that the choice of ESG rating provider has a significant impact on the composition of indices (ESMA, 2021). Companies may or may not qualify for an ESG rating-based index depending on the rating provider, while growing sums of money are being allocated to these indices through passive investment vehicles. The market for ESG ETFs in particular, is highly concentrated: in December 2020, 17 out of the largest 20 ESG ETFs tracked ESG indices from the same index provider (according to data from ETFGI) with combined assets of EUR 57 billion.

The strong growth of passively managed ESG funds, and the prevalence of sustainable investment strategies relying to some extent on ESG ratings within actively managed funds, show that the introduction of transparency rules on ESG fund managers will likely not be sufficient to ensure that sustainable investing makes a significant difference in the long run. The development of a reliable EU-wide label for green funds will help as a first step to identify those funds that are best place to make a positive environmental contribution. For non-labelled funds, the introduction of principal adverse impact metrics (see Chapter 10) should help, in that it will force fund managers to carefully consider their choice of ESG rating provider, and will allow investor to better understand the impact of their investments.

Notes

1. Specifically, the manager is authorized under Directive 2009/65/EC (UCITS) or under Directive 2011/61/EU (AIFMD) or under Directive 2014/65/EU (MiFID II), or under Directive 2009/138/EC (Solvency II) or under Directive (EU) 2016/2341 (IORPs).
2. It is worth noting that, even if the Directive (EU) 2017/828 (Shareholder Rights Directive II) makes a clear distinction between institutional investors (into which insurance undertakings and IORPs fall), on one side, and asset managers (into which UCITS and AIFM entities and MiFID entities providing portfolio management services to investors fall) on the other side, in this paper the term of "institutional investors" is used in a broader sense including any financial managed portfolio regardless of its specific legal nature (collective investment scheme or individual portfolio or insurance portfolio or pension scheme), while "asset manager" refers to the manager of that portfolio. That is in order to provide the reader with a view of characteristics and issues that appear to be common to all managers of ESG products across the different financial compartments.
3. I.e. Prospectus and KIID or KID where relevant.
4. While industry specialists point to differences between SRI and ESG investing (e.g. coverage, strategy type, use of metrics, etc.) the terms are now often used interchangeably.
5. There is no official definition of sustainable investment strategies, therefore alternative classifications exist. For example, EFAMA (2020) distinguish between firm-level and product-level ESG strategies; ESG strategies may further be broken down into sub-strategies, e.g. quantitative, smart-beta, etc. For example, see UN Principles for Responsible Investing (2016).
6. Given the absence of common definitions, it is important to note that some observers categorize "best-in-class" strategies as being part of ESG integration.
7. Directive 2009/65/EC (UCITS) and Directive 2011/61/EU (AIFM). Both UCITS and AIMF Directives lay down a set of rules concerning organizational and behavioural requirements that entities should comply with in order to be authorized under each of them. The main difference between the two frameworks is that, while UCITS Directive defines rules for both managers and products (so that UCITS funds are harmonized across Europe), AIMFD concerns only the manager. The two different approaches, obviously, impact on the passport regime envisaged by the two Directives. Indeed, while UCITS passport is recognized, besides the managers, for the products so that the UCITS funds could be marketed on cross border basis across Europe to both professional and retail investors; the AIFM passport covers only the manager that, on

the basis of passport, can market its AIFs across Member States only to professional investors.

8. Directive 2010/43/EU implementing Directive 2009/65/EC and Regulation (EU) n. 231/2013 supplementing Directive 2011/61/EU.
9. Art. 9 UCITS delegated Directive and art. 60 AIFMR.
10. Art. 23 UCITS delegated Directive and art. 18 AIFMR.
11. Art. 9 UCITS delegated Directive and art. 60 AIFMR.
12. In order to perform a fair and effective decision-making process, the asset manager shall employ personnel with the skills, knowledge and expertise necessary for the discharge of the responsibilities allocated to them (Art. 5 UCITS delegated Directive and art. 22 AIFMR) and—in view to ensure that the decision-making process is appropriately documented to allow its ex post traceability—shall establish, implement and maintain decision-making procedures and an organizational structure which specifies reporting lines and allocates functions and responsibilities clearly and in a documented manner (Art. 4 UCITS delegated Directive and art. 57 AIFMR).
13. Please, see art. 3 g(1), lett. a) of Shareholder Rights Directive II.
14. Recital 14 of Shareholder Rights Directive II.
15. Institutional investors and asset managers shall adopt an engagement policy or, alternatively, publicly disclose a clear and reasoned explanation why they have chosen not to adopt it. The engagement policy shall describe how the institutional investors and asset managers integrate shareholder engagement in their investment strategy, which different engagement activities they choose to carry out and how they do so. In the mentioned context the relevant activities to be considered are, inter alia, the exercise of voting rights and other rights attached to shares, the cooperation with other shareholders and the communication with relevant stakeholders of the investee companies. The engagement policy should also include policies to manage actual or potential conflicts of interests, in particular situations in which the institutional investors or asset managers or their affiliated undertakings have significant business relationships with the investee company.
16. Lastly, it is noticeable that even 2019/2088 Regulation (so-called Sustainability Financial Disclosure Regulation or SFDR: more on this in Chapter 8) provides with an attestation of the relevance of the engagement in the investee companies in the perspective of the enhancement of the inclusion of ESG factors in their business. Indeed, the level 2 measures of SFDR emphasize the role the engagement plays to address principal adverse sustainability impacts and to attain ESG characteristics or objective (accordingly, the ex ante disclosure of the engagement policy and the ex post disclosure of taken actions are required). The recital 15 of the ESAs CD RTS provides even an exemplificative list of the action

that market participants can put in place in relation to principal adverse sustainability impacts (inter alia, "exercising voting rights as a shareholder, sending letters to or attending meetings with the management of investee companies, setting up documented and time-bound engagement in actions or shareholder dialogue with specific sustainability objectives, planning escalation measures in case those objectives are not achieved, including reductions of investments or exclusion decisions").

17. The EU regulatory cycle rests on the Lamfalussy approach, involving four institutional levels. At level 1 the European Parliament and Council adopt the basic laws proposed by the Commission, in the traditional co-decision procedure. As this procedure is usually complex and time-consuming, the Lamfalussy report recommends using it only for setting out framework principles. At level 2 the Commission can adopt, adapt and update technical implementing measures with the help of consultative bodies composed mainly of EU countries representatives. This allows the Council and Parliament to focus on the key political decisions, while technical implementing details can be worked out afterwards by the Commission. At level 3, committees of national supervisors are responsible for advising the Commission in the adoption of level 1 and 2 acts and for issuing guidelines on the implementation of the rules. At level 4 the report advocates a stronger role for the Commission in ensuring the correct enforcement of EU rules by national governments (Source: https://ec.europa.eu/info/business-economy-euro/banking-and-finance/regulatory-process-financial-services/regulatory-process-financial-services_en).
18. CESR Guidelines for the computation of synthetic risk/reward indicator in the KIID UCITS document (CESR/10-673).
19. Specifically: market funds are managed according to investment policies or strategies which aim to reflect the risk and reward profile of some pre-determined segments of the capital market (a benchmark); investment policies/strategies of absolute return funds envisage a variable allocation of the portfolio across asset classes, under the constraint of a predetermined risk limit; investment policies/strategies of total return funds pursue certain reward objectives by participating, through flexible investment, in different financial asset classes (e.g. in both equity and fixed-income markets); life cycle funds are managed according to investment policies/strategies which imply a shifting over time of their portfolio allocation from equity to fixed-income assets as a target maturity date approaches; structured funds provide investors, at certain predetermined dates, with algorithm-based payoffs that are linked to the performance of financial assets, indices or reference portfolios.
20. Delegated Regulation (EU) n. 231/2013—Annex IV that includes the template for the reporting to the Competent Authorities pursuant to art.

24 of AIFMD. Every mentioned macro-category is further detailed in a list of sub categories.
21. ESMA Q&As on application of UCITS Directive (ESMA34-43-392)—Q&A 8b.
22. In the mentioned Q&A ESMA recognizes that the onus of identifying whether the UCITS is in practice managed in reference to a benchmark index is on the UCITS management company.
23. Art. 8 of Disclosure Regulation.
24. Art. 9 of Disclosure Regulation.
25. Art. 15(2) of CD RTS envisages the disclosure of the remainder of the investments other than sustainable investments and other than investments contribute to the attainment of the environmental or social characteristics promoted by the financial product. Moreover, as regards investments that do not qualify as sustainable or as contributing to the environmental or social characteristics, the manager may decide to apply some baseline environmental or social safeguards.
26. When this is the case, the Disclosure Regulation envisages specific obligations concerning—as far as products ex art. 8 are concerned—whether and how the reference benchmark is consistent with the objective characteristics and—as far as products ex art. 9 are concerned—why and how the designated index aligned with that objective differs from a broad market index. These disclosure obligations are further detailed in the Level 2 measures of Disclosure Regulation.
27. Specifically: contributing input data to a benchmark is a voluntary activity, so that all contributors can exercise discretion and are potentially subject to conflicts of interest; the benchmarks are computed on the basis of a formula that are defined, on discretionary basis by the benchmark's administrator. However, it shall be recognized that so-called regulated data benchmark whose input data are not provided by contributors but by regulated and supervised entities (such as trading venues) are less vulnerable to manipulation.
28. Regulation (EU) 2016/1011 on indices used as benchmarks in financial instruments and financial contracts or to measure the performance of investment funds. The regulation is currently under review.
29. Administrators of regulated data are released from certain obligations taking into account that for this kind of benchmarks the risk of manipulation turns to be less relevant.
30. Critical benchmarks—that are the category connoted by the highest manipulation risk—are benchmarks other than regulated-data benchmarks. Critical benchmarks are identified on the basis of a quantitative threshold (benchmark is used directly or indirectly within a combination of benchmarks as a reference for financial instruments, included investment funds,

having a total value of at least EUR 500 billion) or, alternatively, a combination of quantitative and qualitative criteria (the benchmark is based on submissions by contributors the majority of which are located in one Member State and is recognized as being critical in that Member State). The other categories identified by the legislator are significant benchmarks and no-significant benchmarks.

31. Regulation (EU) 2019/2089.
32. Benchmark which is labelled as an EU Climate Transition Benchmark and its underlying assets are selected, weighted or excluded in such a manner that the resulting benchmark portfolio is on a decarbonization trajectory. In addition, these benchmarks shall be constructed in accordance with the minimum standards laid down in the delegated acts.
33. Benchmark which is labelled as an EU Paris-aligned Benchmark and its underlying assets are selected, weighted or excluded in such a manner that the resulting benchmark portfolio's carbon emissions are aligned with the objectives of the Paris Agreement and, moreover, the activities relating to its underlying assets do not significantly harm other environmental, social and governance (ESG) objectives. In addition, these benchmarks shall be constructed in accordance with the minimum standards laid down in the delegated acts.
34. These five eligibility criteria are further complemented by two criteria to disclose information to retail investors and the use of the EU Ecolabel logo. The criteria may be further revised in light of the comments received as part of the latest consultation.
35. For example, the French AMF checks if the information made available to investors is clear, accurate and non-deceptive.
36. This includes, but is not limited to, any reference to investment decision-making or screening based on ESG aspects, reference to ethical investing, socially responsible investing thematic investing, impact investing and engagement with investee companies on ESG-related matters.
37. The difference is even larger for mixed funds (EUR 126 m versus EUR 58 m) but the later average launch date of non-ESG mixed funds means that, for a higher share of these funds, the sample period covers the COVID-19 crisis (when many funds had outflows).

References

Barclays. (2021, April 28). *Hedge fund investor trends—ESG gains traction*. https://www.investmentbank.barclays.com/our-insights/3-point-perspective/esg-gains-traction-among-hedge-fund-investors.html?cid=paidsearch-textads_google_google_themes_hf-esg_we_research_esg_phrase_824445854585&gclid=CjwKCAjw-sqKBhBjEiwAVaQ9a5nFySHZdNJ_jEJbjpASSB3DJ0AkDhIE3uVidKnkZV6NfvxNQ9yH-BoC30gQAvD_BwE&gclsrc=aw.ds

Berg, F., Koelbel, J. F., & Rigobon, R. (2019). *Aggregate confusion: The divergence in ESG ratings* (MIT Sloan Research Papers, 5822-19).

Billio, M., Hristova, I., Latino, C., & Pelizzon, L. (2020). *Inside the ESG ratings: (Dis)agreement and performance* (Working Papers from Department of Economics, University of Venice Ca' Foscari, No. 2020:17). https://EconPapers.repec.org/RePEc:ven:wpaper:2020:17

Boffo, R., Marshall, C., & Patalano, R. (2020). *ESG investing: Environmental pillar scoring and reporting*. OECD.

Boffo, R., & Patalano, R. (2020). *ESG investing: Practices, progress and challenges*. OECD.

CFA Institute. (2015). *Guide to ESG investing for investment professionals*.

Eccles, N., & Viviers, S. (2011). The origins and meanings of names describing investment practices that integrate a consideration of ESG issues in the academic literature. *Journal of Business Ethics, 104*, 389–402.

EFAMA. (2020). Sustainable investment in the European asset management industry. *EFAMA Market Insights*, 3. https://www.efama.org/sites/default/files/publications/Market%20Insights%20Issue3%20ESG%20strategies_1.pdf

ESMA. (2021). *ESG ratings—Status and key issues ahead* (Report on Trends, Risks and Vulnerabilities, 1).

European Commission, Joint Research Center. (2021, March). *Development of EU Ecolabel criteria for retail financial products* (Technical Report 4.0). https://susproc.jrc.ec.europa.eu/product-bureau//sites/default/files/2021-03/2021.03.05%20-%20EUEL%20financial%20products%20-%20Technical%20Report%204%20FINAL.pdf

Eurosif. (2018). *European SRI study 2018*. Retrieved from https://www.eurosif.org/wp-content/uploads/2018/11/European-SRI-2018-Study.pdf.

Freshfields Bruckhaus Deringer. (2005). *A legal framework for the integration of environmental, social and governance issues into institutional investment*.

Gray, J. (2012). Misadventures of an irresponsible investors. *Rotman International Journal of Pension Management, 5*, 8–13.

J.P. Morgan (2020, May 18). *ESG and COVID-19: Friends or foes? J.P. Morgan perspectives*.

Pew Research Center. (2021, May 5). *Millennial and Gen Z Republicans stand out from their elders on climate and energy issues*. https://www.pewresearch.org/fact-tank/2020/06/24/millennial-and-gen-z-republicans-stand-out-from-their-elders-on-climate-and-energy-issues/

PRI et al. (2015). *Fiduciary duty in the 21th century*.

SASB (Sustainability Accounting Standards Board). (2018). *SASB Standards Application Guidance*.

Sparkes, R. (2001). Ethical investment: Whose ethics, which investment?. *Business Ethics: A European Review, 10*, 194–205.

UN Principles for Responsible Investing. (2016). *A practical guide to ESG integration for equity investing*. Retrieved from https://www.unpri.org/listed-equity/a-practical-guide-to-esg-integration-for-equity-investing/10.article

CHAPTER 7

Retail Investors' Attitude and Preferences and Sustainable Investing Regulation

Barbara Alemanni

Nothing is as powerful as an idea whose time has come. –Victor Hugo

Introduction

Sustainable investments have been gathering more and more attention over the years and, according to the already mentioned GSIA estimates (GSIA, 2020; see also Chapter 2), at the start of 2020 global sustainable investments reached $35.3 trillion, representing a 36% increase in two years and 2.8 times the outstanding amount in 2012.[1] Although institutional investors tend to dominate the financial market, interest by retail investors in sustainable and responsible investing has been steadily

B. Alemanni (✉)
Bocconi University, Milan, Italy
e-mail: barbara.alemanni@unibocconi.it

University of Genoa, Genoa, Italy

N. Linciano et al. (eds.), *Information as a Driver of Sustainable Finance*, Palgrave Studies in Impact Finance,
https://doi.org/10.1007/978-3-030-93768-3_7

growing since statistics began differentiating between these two investment classifications in 2012. At that time, institutional investors held 89 per cent of assets compared with 11 per cent held by retail investors. At the start of 2018, the retail portion had grown to one quarter.

This chapter intends to investigate how retail investors approach sustainable investments and how the regulatory framework in the field is evolving. Section "Retail Investor Attitude Towards Sustainable Investments" introduces the literature shading light on the motivation for households to invest in a social responsible and more generally sustainable way, on the characteristics of consumers willing to take into consideration sustainable factors when investing and, in the end, on the justification for an attitude-behaviour gap regarding sustainable investments. Section "EU Regulation and Retail Investors' Sustainable Investments" explores the general regulatory framework designed to protect retail investors, discusses the limitation of full disclosure as a protection tool and the need for other more paternalistic measures. Section "Concluding Remarks" concludes with remarks on retail investors' heterogeneous preferences for ESG and how transparency and conduct of business can accommodate for it.

Retail Investor Attitude Towards Sustainable Investments

Motivation of Responsible Investing

The rationale behind the introduction of more sustainable financial products initiates in the observation that consumers are increasingly buying socially responsible products. In 2018, the market share of green energy in EU reached a record level representing almost 20% of total consumption. The UK Fair-trade foundation claims to have labelled over 4500 products as Fair trade in 2020. From 2009 onwards, more than 100 million pounds of Fair Trade certified coffee was imported into the US alone of which 62% was organic. These numbers reveal non-ignorable preferences towards sustainable consumer product attributes. Nevertheless, in the investment field, the classical economic literature focuses on risk and return. Utility from non-monetary aspects is largely left out of consideration. What the classical approach is missing out is that most individual investors consider investments as consumption goods and not just as investment products (Keloharju et al., 2012), meaning that non-financial attributes can influence the decision to invest or not.

Most research finds that retail investors focus not only on investments' risk-return ratio but also on the moral or value expressive benefits that increase their utility when making sustainable investment decisions (Beal et al., 2005; Hofmann et al., 2008; Hong & Kacperczyk, 2009; Martin & Moser, 2016; Statman, 2008). Sparkes and Cowton (2004) identify three possible motivations for investors to pursue sustainable investments. Firstly, investors may be motivated to invest in line with their own beliefs and values. Within this group there are "feel good" investors, who feel better about making responsible investments, and there are investors who are motivated by pursuing social change (Lewis, 2001; Schueth, 2003), with others driven by various levels of religiosity (Tahir & Brimble, 2011). But why do investors benefit from moral or value expressive attributes? From an economic perspective, altruism serves to explain such utility as an exception to rational behaviour. Research demonstrates that people might have impurely altruistic motives and receive additional utility from the mere act of being prosocial (Andreoni, 1989, 1990). This "warm glow" effect can be explained by people caring about who they are (Bénabou & Tirole, 2006, 2011). In the case of valuing corporate social responsibility (CSR), Friedman and Heinle (2016) model such "warm glow" related to being an owner of a firm with a positive externality. When investors gain utility from CSR based on preferences for a nonfinancial output, maximizing shareholder value is not equivalent to maximizing the discounted value of cash flows. Additionally, Heeb et al. (2021) find out that investors experience positive emotions when choosing a sustainable investment, irrespective of the investment's impact and conclude that the average prosocial investor can be viewed as a warm glow optimizer, rather than a consequentialist who optimizes the impact of his/her investments.

A second motivation is that of the "rational investor" seeking higher returns. These investors perceive that there is (or could be) a performance premium to be gained from investing in companies with high sustainability metrics and are not particularly aligned with any moral philosophy (Williams, 2007).

The third motivation comes from a social activist standpoint. Investors in this group invest in a company in order to change its behaviour. The shareholders' activism is manifested through letter-writing campaigns, divestment, dialogue with corporate leaders and their submission of resolutions at the company's annual meetings. Thus shareholder advocacy is a form of social movement activism that seeks changes through direct communication with the management in corporate social policy and practice (King & Pearce, 2010). It is different from other social movement activism, in that most participants are investors within the companies they seek to change. The shareholder activists' aim is to enhance the well-being of all stakeholders, including other share owners, customers, employees, vendors, communities and the natural environment.

Different empirical approaches support the notion of investors' nonfinancial utility. A large strand of literature has empirically investigated money flows into and out of sustainable funds (e.g. Benson & Humphrey, 2008; Bollen, 2007; Peifer, 2014; Renneboog et al., 2011). Collectively, the studies suggest that investors derive nonfinancial utility from sustainable fund attributes. For example, investors' cash flows are significantly less sensitive to past negative returns than their conventional fund flows. Focusing on institutional investors, Petersen and Vredenburg (2009) explore this question through both qualitative interview-based research with corporate executives and institutional investors and quantitative survey-based research with institutional investors. They find that most institutional investors were not willing to pay a premium for shares of sustainable companies that engaged in CSR but did favour holding shares in those companies. As disentangling the investor motives is difficult with archival data, Martin and Moser (2016) explore this issue with an experimental study, which allows to isolate the value investors put on the societal benefits of green investments. They find that investors' positive reactions to social responsible information are driven by the disclosed societal benefits of the investment rather than by the disclosed cost to the company. This finding is in line with those of the experimental studies of Barreda-Tarrazona et al. (2011) and Pasewark and Riley (2010), which provide evidence of investors' concerns about the societal effects of their investment decisions and investors' benefits from expressing their preferences for sustainable investments. Barreda-Tarrazona et al.'s (2011) results suggest that in addition to incorporating risk-return characteristics and diversification criteria in their investment decisions, participants invest significantly more in a fund when they are explicitly told about its sustainable nature. Moreover, Rubaltelli et al. (2015) report experimental evidence on nonfinancial motivations influencing investors' reactions to an unsatisfactory investment performance. A recent study that links administrative investor data to survey responses and behaviour in incentivized experiments finds that investors with a strong social motivation are even willing to forgo financial returns in order to invest in accordance with their social motives (Riedl & Smeets, 2017).

Box 7.1: SRI funds performance
The first analysis of socially responsible fund performance was conducted by Moskowitz (1972). Since then, a substantial number of articles have been published on SRI portfolio performance. The extant literature has mixed findings regarding the performance and resilience of socially responsible funds compared to conventional funds (see also Chapter 6 for a review of the literature). Some authors, for example, Michelson et al. (2004), Kempf and Osthoff (2007), Gil-Bazo et al. (2010) and Duranda et al. (2013) view socially responsible firms as performing better than conventional ones due to social screening practices, sustainability, management quality, longer time horizon operation, less agency cost and optimal financial policies of the socially responsible firms. Some empirical investigation, for example, Renneboog et al. (2008), indicate that conventional funds outperform SRI stocks in the European market. From the experience of many European, North American and Pacific Asian markets, the study finds that there is a cost to being socially conscious or ethical. However, this same study shows that socially responsible funds outperform conventional funds in Sweden, Japan and France due to differences in risk-adjusted returns of socially responsible funds. Lee et al. (2010) highlight that conventional funds perform better than socially responsible funds. However, a large number of studies (Bauer et al., 2005, 2006; Becchetti et al., 2015; Duranda et al., 2013 among others) indicate no significant differences in portfolio performance between SRI and conventional funds. Moreover, according to Almazan et al. (2004), no significant difference has been found in return performance of SRI and conventional firms after controlling fund size, portfolio turnover policy, restrictions or constraints. More recent empirical studies, investigating the performance of sustainable funds, confirm such findings. For example, Morgan Stanley (2019), conducting a research on the performance of nearly 11,000 mutual funds from 2004 to 2018, shows that there is no financial trade-off in the returns of sustainable funds compared to traditional funds; additionally, sustainable funds demonstrate lower downside risk. Arefeen and Shimada (2020) document a higher resilience of Japanese sustainable funds during shock periods such as the 2016 American elections and Brexit.

Individual Characteristics

Besides motivation, another important area of investigation deals with the influence of individual determinants on the demand for sustainable investments and on the investor evaluation of sustainability-related information. Many studies have considered the demand side of SRI, particularly investor demographics as an explanatory variable for SRI preference. The results of these studies are varied. For example, Schueth (2003), Tippet and Leung (2001), Nilsson (2008) and Junkus and Berry (2010) all find gender to be significant, with females more likely to be SRI investors; however, Woodward (2000) does not find gender to be significant. Education has also been linked to SRI (Schueth, 2003). Rosen et al. (1991) and Tippet and Leung (2001) find evidence to support SRI investors being better educated, and they also find that SRI investors tend to be younger. In contrast, McLachlan and Gardner (2004) show no evidence that SRI investors are younger or better educated. Finally, an Italian study (Linciano et al., 2019) shows that wealth, financial knowledge, social preferences, climate concerns and consideration of ESG factors as well as tolerance to short-term and tolerance to small losses are among the factors positively associated with familiarity and interest in SRIs, while risk aversion and loss aversion are among the variables showing a negative correlation.

No matter the empirical data, there is strong support for an increasing demand for sustainable investments among young generations. The Millennial generation, also called generation Y (birth years: 1981–1996), will play a significant role in creating demand for sustainable financial products and services. Their interest in this new way of investing is driven by their desire to invest in line with their values. Indeed, according to a survey by Morgan Stanley, millennials are twice as likely than other investors to invest in companies with significant positive social or environmental impacts (Morgan Stanley, 2017). This is especially important due to the ongoing intergenerational wealth transfer to Millennials, with Deloitte's projections suggesting an over twofold increase from 2015 to 2020 levels in this generation's aggregate net worth, estimated to reach USD 19–24 trillion (Kobler et al., 2015).

Millennials might indeed play an important role in increasing demand for sustainable investing, as surveys suggest. At the same time, it remains to be seen what percentage of survey respondents that indicate their preferences in favour of ESG will actually give effect to their will. In other words, the adoption of sustainable investing by millennials might be slower than these surveys seem to suggest. In this respect, Millennials are as likely as other generations to be caught by the attitude-behaviour gap analysed in the following subsection.

Investors' characteristics also appear to play a role in the way sustainability-related information is evaluated. Similar to heterogeneity in conventional investing, common individual determinants influencing investor evaluation include investment practice and skill, investment horizon, investor experience, financial literacy (Borgers & Pownall, 2014; Holm & Rikhardsson, 2008; Jain et al., 2016), and socio-demographic factors (Cheah et al., 2011; Säve-Söderbergh, 2010). Beyond these common factors, individual determinants influencing investors' evaluation of sustainability-related information mainly refer to investors' social preferences and identification. Nilsson (2008, 2009) and Jansson and Biel (2014) provide evidence that investors' attitudes towards sustainability affect investment decisions in the context of sustainable investing. Research also demonstrates heterogeneity in investment decisions due to individuals' level of concern about sustainability issues (Barreda-Tarrazona et al., 2011; Cox et al., 2004; Renneboog et al., 2011). Additionally, investment in socially responsible assets is higher for investors who perceive higher levels of moral intensity in the situation of decision-making (McLachlan & Gardner, 2004; Webley et al., 2001). In particular, Bauer and Smeets (2015) find that social identification in a relevant social group influences investors' attitude. Survey data from retail investors of two socially responsible banks show that investors with a strong social identification allocate substantially more of their wealth to these banks. Social identification also mediates the effect of expected return on SRI. Furthermore, Riedl and Smeets (2017) discover that investors' social preferences are a major factor determining the likelihood of holding socially responsible equity funds. In addition, Glac (2009, 2012) examines the role of mental frames in investors' decision-making processes and finds that investors' mind-sets influence both the likelihood that they will choose an SRI over a non-SRI fund option (Glac, 2009) and the percentage of SRI assets in their portfolios (Glac, 2012).

The Gap Between Self-Reported Attitude and Actual Behaviour

A final strand of relevant literature refers to the investors' actual behaviour. Similarly to what it is registered for sustainable goods and products where consumers declare an interest in sustainability but fall short in their actual purchasing behaviour—an attitude known as "attitude – behaviour-gap" or "value-action gap" (Boulstridge & Carrigan, 2000), investors do not invest in sustainable assets as much as they proclaim. According to the 2019 Schroders Global Investment Study, the proportion of people who are investing sustainably is significantly lagging those who are interested and would like to invest, indicating that a gap exists between investors' intentions and their tangible actions. The study, which garnered the views of more than 25,000 investors in 32 locations around the world, identified that 16% of people invest in sustainability, compared to 32% who say they are interested and would like to invest this way. A similar gap between motivation and action is documented for Australian superannuation members: de Zwaan et al. (2015) find that Australian savers bear a remarkable interest in sustainable investments, but only occasionally they choose to allocate their pension saving to sustainable products. They lack awareness of their fund's approach to ESG investing. Limited information could explain Schroders results as well. In fact, the research found that a majority of investors believe fund managers should be automatically investing sustainably. Nearly two-thirds (61%) of investors thought all investment funds should consider sustainability factors, not just those specifically designed as "sustainable investment funds".

The attitude-behaviour gap is documented in Italy as well. A recent survey by CONSOB (Linciano et al., 2019) shows that 40% interviewees declare a potential interest in SRIs, but only 5% actually hold them, the proportion rises to 18% among informed advised investors, who report to have been recommended such investments by their advisors in slightly more than 10% of the cases. Additionally, the survey informs that the main deterrents from interest in SRIs seem to be lack of savings and mistrust, the latter entailing several dimensions as greenwashing' concerns, ineffectiveness of SRIs and inclination towards keeping personal engagement separate from financial choices.

The reasons behind the value-action gap in sustainable consumption have been largely investigated (e.g. Hirsch & Terlau, 2015) and find roots in individual, social and situational factors. Among others cognitive factors appear to play an important role in explaining the gap and better way to inform consumers are under the lenses of both the policy makers and the industry. Information and communication are increasingly an important part of research in the field of responsible investment as well.

Barreda-Tarrazona et al. (2011) indicate that, beyond return and diversification, investors who are informed of the socially responsible nature of the funds invest more. Statman (2008) argues that it is the role of the advisor to explore social preferences, ethical positions and religious beliefs in the same way he explores attitudes vis-à-vis risk. In an empirical study, Diouf et al. (2016) find that investors advised on SRI are more likely to demand them.

Young et al. (2010) address the role that information can play. Indeed, many clients express their desire to take an interest in ethical products but find it difficult to put this into practice and the research shows that it is crucial to establish incentives and even labels to accompany the efforts of consumers. Eco-labels, or in the case of financial assets, ESG rating, are important instruments to reduce the cognitive burden savers face, but they can be double-edged swords as Rzeźnik et al. (2021) show. The inversion of the rating scale led some investors to make incorrect assessments about the meaning of the change in ESG ratings. They bought (sold) stocks they misconceived as ESG upgraded (downgraded) even when the opposite was true (Table 7.1).

EU Regulation and Retail Investors' Sustainable Investments

Private sector savings in Europe amount to some 20% of GDP—a valuable asset, if efficiently used, to stimulate growth and job-creation and more recently to foster the transition to a more sustainable economy. As it will be discussed in Chapter 8, in March 2018, the European Commission published an ambitious Action Plan on Sustainable Finance, which proved to be the first step in a series of regulatory reforms aimed at fundamentally reorienting capital flows towards sustainable investment and managing perceived financial risks stemming from climate change. In such a scenario, the retail investor is profiled as a saver but also as a capital-supplier and risk-taker. This suggests a regulatory regime focused on engagement with the markets, choice, empowerment, removing barriers to market access and improving awareness.

Table 7.1 Research areas regarding retail investors' attitude and characteristics and sustainable investments

Motivation of investments	Non-monetary utility: pure and impure altruism	Beal et al. (2005), Hofmann et al. (2008), Hong and Kacperczyk (2009), Martin and Moser (2016), Statman (2008), Lewis (2001), Schueth (2003), Tahir and Brimble (2011), Andreoni (1989, 1990), Bénabou and Tirole (2006, 2011), Friedman and Heinle (2016), Heeb et al. (2021)
	Empirical evidence of non-monetary utility	Benson and Humphrey (2008), Bollen (2007), Peifer (2014), Renneboog et al. (2011), Petersen and Vredenburg (2009), Martin and Moser (2016), Barreda-Tarrazona et al. (2011), Pasewark and Riley (2010), Rubaltelli et al. (2015), Riedl and Smeets (2017)
	Optimization of risk and return	Williams (2007)
	Social activism	King and Pearce (2010)
Investors' characteristics	Socio-demographic characteristics and demand for sustainable products	Schueth (2003), Tippet and Leung (2001), Nilsson (2008), Junkus and Berry (2010), Woodward (2000), Schueth (2003), Rosen et al. (1991), Tippet and Leung (2001), McLachlan and Gardner (2004), Linciano et al. (2019)
	Individual determinants and investment evaluation	Borgers and Pownall (2014), Holm and Rikhardsson (2008), Jain et al. (2016), Cheah et al. (2011); Säve-Söderbergh (2010), Nilsson (2008, 2009), Jansson and Biel (2014), Barreda-Tarrazona et al. (2011), Cox et al. (2004), Renneboog et al. (2011), McLachlan and Gardner (2004), Webley et al. (2001), Bauer and Smeets (2015), Riedl and Smeets (2017), Glac (2009, 2012)

(continued)

Table 7.1 (continued)

Actual investors' behaviour and limit to sustainable investments	Attitude-behaviour gap	Boulstridge and Carrigan (2000), de Zwaan et al. (2015), Linciano et al. (2019), Hirsch and Terlau (2015), Young et al. (2010)
	The role of information and communication	Barreda-Tarrazona et al. (2011), Statman (2008), Diouf et al. (2016), Rzeźnik et al. (2021)

A General Framework of Retail Investors' Regulation

Attempting to build a retail investor culture through law represents a major challenge in EU policy. It reflects an international policy concern with individual financial literacy and independence, with promoting greater diversity in personal saving mechanisms and with seeking to understand how retail investors behave.

While protection and confidence-generation is the dominant function, regulation also has a more proactive role in retail policy. It can support investor empowerment, investor choice and the ability of investors to demand appropriate products and services.

The twin pillars of the European Union's retail strategy, investor protection and investor choice, meet in the conduct of the business regime. Measures aimed at improving disclosure and transparency reflect investor protection objectives, given the informational asymmetries in the firm–investor relationship. But they also suggest a strong characterization of the investor as rational agent. Rational and self-disciplined wealth maximizing investors need large volumes of information in order to calculate the risk and return probabilities of an investment, in order to maximize their expected utility (expected profit) in accordance with their risk and return preferences. However, the evidence suggests that investors make poor decisions, ignore relevant information and rely on biases and mental short cuts. At the roots of behavioural approach to choice is Kahneman and Tversky (1979), and Tversky and Kahneman (1992) Prospect Theory which assumes, inter alia, that preferences are not constant and choices may be manipulated through the framing of information.

Disclosure seems less effective as a regulatory technique, if the effects of problem description or framing are also taken into account. There is strong evidence that preferences may change, depending on the wording of a problem. This essentially means that individuals' choices can be manipulated depending on the way relevant information is presented. As LeBouf and Shafir (2003) document, the effect of framing is stronger among the less sophisticated members of any group. However, even thoughtfulness is not sufficient to counter the effect of framing; thoughtful individuals are still in need of a relevant cue in order to untangle the impact of framing. In addition to limitations in the use of information highlighted by Prospect Theory, there are other factors that seem to limit the effectiveness of disclosure. A relevant aspect has to do with what Camerer et al. (1989) describe as the "curse of knowledge", which is the information overload occurred by investors. More generally, we commonly think of information as a means to an end, however, as Golman et al. (2017) document, there is a growing theoretical and experimental literature suggesting that information may directly enter the agent's utility function.[2] This can create an incentive to avoid information, even when it is useful, free and independent of strategic considerations.

Additionally, in the case of disclosed information regarding highly complex financial instruments, investors might be unable to use and understand them due to what Simon (1955) defined as "bounded rationality", the limited ability to process information because of limited computational ability and flawed memory. Disclosure could also be ineffective because of herding, when individuals ignore disclosed information in favour of the safer "follow the herd strategy".

Behavioural studies point out that investors' ability to acquire and employ specific information highly depends on the extent to which they attend to this information, as they cannot process all available information (Barber & Odean, 2008; Da et al., 2011). In this context, the salience of information and the ease with which related memories are accessed help drive information acquisition (Bordalo et al., 2013; Hirshleifer & Teoh, 2003). Whether an investor acquires a specific piece of information also depends on characteristics of the specific item and the information environment.

In the case of sustainable finance, the limited function of disclosure might appear particularly severe. Corporate provision of non-financial information, such as CSR-related information, in addition to the existing financial statement may lead to information overload for stock market participants (Neumann et al., 2011). Additionally, given that attention is a scarce cognitive resource (Kahneman, 1973), research also indicates that attention constraints can reduce investors' choice sets by distracting attention from CSR information. Pilaj (2017) proposes that potential low salience of CSR information and, thus, limited investor attention to this information can reduce engagement in sustainable investment practices. Modelling investor decision-making in the context of socially responsible investing (SRI), Pilaj (2017) argues that limited attention at the moment of choice can lead to unintended disregard of CSR decision criteria when investing.

Not all evidences agree on a low salience of non-financial information. Event studies show that indeed investors attend and react to CSR announcements (Flammer, 2015; Jones & Murrell, 2001). In addition, Wang et al. (2011) suggest that events or characteristics that direct investors' attention to CSR issues can drive sustainable investment behaviour in the market. Analysing abnormal returns and abnormal net order flow at the firm level before and after the melamine contamination incident in China, they report that in the post-event period, firms' disclosed CSR information significantly influenced both individual and institutional investor behaviour, whereas this was not the case before the incident.

Since disclosure may not protect investors if cognitive biases and other attitudes prevent them from rationally incorporating the information disclosed into their investment decisions and given the limitations in decision-making due to bounded rationality, regulatory concern to protect but also to enhance the quality of choices should be resolved through the adoption of other strategies. Defects in investor decision-making and information deficiencies, which can deter investors and damage the reputation of market investment, can be countered by buying in expertise through intermediation and investment advice. However, this investor-investment firm relationship generates risks, additional to those of market investing, which may call for intervention. It is characterized by informational asymmetries and once again imperfect information can expose investors to risk with respect to fraud, misuse of funds, conflicts of interest and the competence of the firm in advising on and executing transactions. Regulation therefore imposes disclosure standards on the investor–investment firm relationship and it extends to more general conduct of business rules. It intervenes in the management of this relationship given that reliance on an investment firm, particularly where investment advice and portfolio management are concerned, sees the investor placing trust in the reputation, expertise and competence of the firm and, in essence, in its gatekeeper function. Suitability/"know-your-client" requirements are traditionally associated with a more paternalistic, managerial approach to the retail investor.[3] They reflect the information asymmetry in the firm–investor relationship and the reliance of the investor on the investment firm's expertise and judgement. Increasingly, they are associated with minimizing the behavioural defects under which investors operate.

The suitability regime does, therefore, intervene in the investor–investment firm relationship by imposing proactive, and individualized, suitability obligations on firms, considerably beyond the broadly standardized disclosure and contractual requirements.

To reinforce further the protection for investors with limited cognitive abilities, EU regulators introduce a product governance rule which should prevent that products are sold or recommended to investors for whom they are not intended and come on top of pre-existing measures, such as the suitability and appropriateness assessments. Product governance is a kind of "know-your-client" requirements at product level: both in the design and the distribution phase of a product, investment firms should identify the target market for each product. As a result, product manufacturers and distributors need to properly understand all product features and assess with what type of clients those products would be compatible.

Targeting Retail Investors to ESG Investments in the New EU Regulation

The EU has set itself a major challenge of re-orientating investor capital towards sustainable investment, given the ambitious targets it has set in the Action Plan. Through the latter, the EU has set out to build an ambitious and comprehensive agenda to incorporate concepts of sustainability into the existing regulatory approach for financial markets. The combined elements of the proposed suite of measures look at sustainability through the lens of conduct, transparency, investor protection and even from a prudential perspective.

The Sustainable Finance Disclosure Regulation (SFDR) aims at creating a common language on sustainability matters in the relationship among financial firms and investors. It is part of a raft of legislation published as part of the Action Plan on sustainable finance in March 2018. As mentioned in Chapter 4, this includes a Regulation amending the Benchmarks Regulation in relation to low carbon benchmarks and positive carbon impact benchmarks; a Regulation on the establishment of a framework to facilitate sustainable investment (the Taxonomy Regulation). Additionally in April 2021, within an ambitious and comprehensive package of measures to help improve the flow of money towards sustainable activities across the European Union, the Commission also adopted six amending Delegated Acts on fiduciary duties, investment and insurance advice that will ensure that financial firms, e.g. advisers, asset managers or insurers, include sustainability in their procedures and their investment advice to clients.[4] For further details and considerations about potential issues raised by the application of these rules, see Chapter 8.

Concluding Remarks

If the common man derives positive utility from more socially responsible investments, asset managers should incorporate these preferences of their clients into their investment schemes. However, the approach to including a sustainability objective takes different forms for different investors. For some, it is a dedicated effort to avoid exposures and risks. For others, it is a dedicated effort to advance a certain objective. Each of these motivations are legitimate and should be met through an approach broad enough to capture them. This is not an easy task, Barber (2007) points out: "*Once considerations other than wealth maximization are relevant for investors, aligning the interests of portfolio managers and investors becomes extremely difficult*". Think about regulatory difficulties concerning heterogeneous preferences (Richardson, 2011) or the interpretation of fiduciary duties (Richardson, 2007, 2009). However, these difficulties cannot be an argument to refrain from taking environmental and social preferences into account at all and represent them effectively to the public.

A key ingredient in this effort will be achieving a common understanding—across investors, asset managers, other market participants and regulators—of what is expected from financial products that offer exposure to sustainable investment themes. This requires a strong system of classification—in regulatory parlance a "taxonomy"—that enables investors to differentiate products and provides clear, transparent data regarding product attributes. The EU Taxonomy regulation is meant to meet this need (see Chapter 10).

The policy can and should allow—and even encourage—a diversity of sustainable products, providing a range of different approaches to sustainable investment to suit an array of investors' needs and motivations. This is best accomplished by having a common lexicon of sustainable investment approaches, underpinned by transparent data at the product or portfolio level. This enables both regulators and investors to best understand and assess the connection between the product label and what is in the product. Maintaining a healthy variety of different approaches to investing around a theme of sustainability is key to facilitating investor choice, which is in turn critical to underpin the continued growth in this space.

This focus on clear, objective data can help investors better understand a range of different sustainable investment strategies and help asset managers describe their approaches and measure performance. The focus moves from subjective determinations of what is and isn't sustainable, with the attendant risks of greenwashing, and, instead, provides investors the necessary data for them to determine the product's alignment with their goals relative to other investment options. All this means building a policy that requires asset managers and product providers to describe their investment approaches clearly and understandably to investors, regulators and to other stakeholders. Language should explain what a fund means when it describes how sustainability-related concepts factor into its investment process and ensure investors have meaningful information to understand how that translates into their investment.

However, the risk for consumers to fall into an information overload is far from remote and other initiatives must match these classification and disclosure efforts. For instance, in July 2021 the Commission published a proposal on an EU Green Bond Standard (GBS; see also Chapter 4) and is considering the introduction of an eco-label[5] to encourage retail savers to buy green investments and it would be useful that similar standards should be developed for social bonds as well. Similar initiatives proved to be useful in consumer goods. Even if consumers lack the technical knowledge to understand the energy efficiency of electronic equipment, they are able to select on the basis of a simple and straightforward eco-labelling.

A second key effort and challenge regards the conduct of business of both advisors and manufacturers and refers to effectively embedding ESG considerations within the suitability process and the product governance.

Many may argue that suitability requirements are already overly burdensome for clients, and adding further requirements on ESG preferences risk making the process overly lengthy and complex, ultimately jeopardizing customer engagement. It is therefore important that firms consider the customer journey in its entirety.

Before asking a client what their ESG preferences are, advisors will first need to ensure clients understand what is meant by ESG in clear and simple language. This may include describing the difference between the "E", the "S" and the "G" and what sort of factors are considered under each heading. In the same way advisors have to assess clients' knowledge, experience and understanding of investment risks associated with the advised products, firms may wish to consider whether they should do the same for ESG. For example, a client who misinterprets ESG might limit what products the advisor considers for them. In future years, if the advisors' recommendation has not achieved the desired outcome for the client, it could give rise to a suitability-related complaint.

Client responses could fall across a spectrum of attitudes and preferences for ESG that advisors must reconcile and consider. This could range from clients having no ESG preference to clients having very specific preferences towards achieving an impact that aligns with a particular Sustainable Development Goal (SDG), otherwise known as Impact Investing.

Advisors also need to be aware that people's attitudes to ESG factors may change over time. Investors' preferences in response to world events and topical issues at the time of assessment will likely influence their responses. Striking the right level of granularity in the questions asked will therefore be key, too niche or too detailed could limit investment choice or lead to portfolio rebalances over time; too high level and firms risk not capturing an investor's actual ESG preferences.

When making a recommendation, advisors need to be clear how they have considered the client's ESG preferences for each financial instrument recommended. In particular, suitability letters should contain an explanation on whether, or how, clients' objectives have been achieved by taking into account their expressed ESG preferences. To do this, advisors need to have clear consistency between the internal categorization of ESG products and external marketing information for ESG products. Embedding the suitability process with product governance oversight is a key step to implement.

An additional key concern is around fee transparency and the potential conflicts of interest that could exist between firms and clients. Some ESG products are, by nature, costlier than their non-ESG equivalents and this could create conflicts where funds or strategies are labelled as ESG in order to charge higher fees. To this end, ESMA set out in its technical advice that firms should have in place appropriate arrangements to ensure that the inclusion of ESG considerations in the advisory process does not lead to mis-selling practices or misrepresentations and do not damage the interest of the client. Or, as ESMA describe it, "an excuse [for a firm] to sell its own-products or more costly ones, or to generate unnecessary churning of clients' portfolios, or by firms misrepresenting products or strategies as fulfilling ESG preferences where they do not".

Firms will need to consider any limitations of how particular ESG preferences may affect uptake of model portfolios and investment strategies, and, ultimately whether the clients' objectives can be met where the client has expressed strong ESG preferences. But the more fundamental question is the trickiest to solve, because it boils down to ethics rather than finance. How can the relative value of, say, educating a girl in the developing world be compared with preventing a tonne of air pollution? In the end, investors' choices among the different variants of sustainable investments will be driven by their own personal interests, rather than just by financial calculations.

Notes

1. The Global Sustainable Investment Alliance (GSIA, 2019) defines sustainable investing, also called responsible investing or values-based investing, a way to invest that involves incorporating environmental, social and governance factors when making investments decisions rather than relying purely on financial consideration (see this chapter for alternative definitions). The GSIA definitions of sustainable investment, published in the Global Sustainable Investment Review 2012, encompass several investment strategies, from positive/negative screening to ESG integration, and have emerged as a global standard of classification.
2. Although people often fail to collect or attend to potentially helpful information, only a small fraction of such instances qualify as active avoidance. Golman et al. (2017) specify two necessary criteria for avoidance to be classified as "active": (1) the individual is aware that the information is available, and (2) the individual has free access to the information or would avoid the information even if access were free. The reasons why people

avoid information can be consistent with standard economic theory, broadly construed. For example, information can be avoided for strategic reasons; ignorance can be a very useful commitment device (Schelling, 1960). Other reasons are inconsistent, or less consistent, with standard theory. Most importantly, information can have direct hedonic value (positive or negative), separate from its usefulness. Behavioral theories that incorporate such "belief-based utility" can help explain why a person might avoid information that has material value.

3. According to the Directive 2014/65/EU (Markets in financial instruments directive or MiFID II), intermediaries providing investment services such as investment advice or discretionary portfolio management have to perform the "suitability assessment", i.e. collect information about clients on their objectives, knowledge and experience, and financial situation, and subsequently assess that a given investment product is suitable for them, based also on the firm's solid understanding of the products that it can recommend or invest into on behalf of the client (see also ESMA Guidelines on certain aspects of the MiFID II suitability requirements, https://www.esma.europa.eu/sites/default/files/library/esma35-43-1163_guidelines_on_certain_aspects_of_mifid_ii_suitability_requirements_0.pdf).
4. The EU Ecolabel for Retail Financial Products is being developed by Unit B5—Circular Economy and Industrial Leadership, as well as Unit B1—Finance & Economy of the Joint Research Centre (JRC) Directorate B—Growth and Innovation for the Directorate General for the Environment in collaboration with the Directorate General for Financial Stability, Financial Services and Capital Markets Union (DG FISMA) of the European Commission.

References

Almazan, A., Brown, K., Carlson, M., & Chapman, D. (2004). Why constrain your mutual fund manager? *Journal of Financial Economics, 73*, 289–321.

Andreoni, J. (1989). Giving with impure altruism: Applications to charity and Ricardian equivalence. *Journal of Political Economy, 97*, 1447–1458.

Andreoni, J. (1990). Impure altruism and donations to public goods: A theory of warm-glow giving. *Economic Journal, 100*, 464–477.

Arefeen, S., & Shimada, K. (2020). Performance and resilience of socially responsible investing (SRI) and conventional funds during different shocks in 2016: Evidence from Japan. *Sustainability, 12*, 1–20.

Barber, B. M. (2007). Monitoring the monitor: Evaluating CalPERS' activism. *Journal of Investing, 16*, 66–80.

Barber, B. M., & Odean, T. (2008). All that glitters: The effect of attention and news on the buying behavior of individual and institutional investors. *Review of Financial Studies, 21*, 785–818.

Barreda-Tarrazona, I., Matallin-Saez, J. C., & Balaguer-Franch, R. M. (2011). Measuring investors' socially responsible preferences in mutual funds. *Journal of Business Ethics, 103*, 305–330.

Bauer, R., Kees, K., & Otten, R. (2005). International evidence on ethical mutual fund performance and investment style. *Journal of Banking and Finance, 29*, 1751–1767.

Bauer, R., Otten, R., & Rad, A. T. (2006). Ethical investing in Australia: Is there a financial penalty? *Pacific-Basin Financial Journal, 14*, 33–48.

Bauer, R., & Smeets, P. (2015). Social identification and investment decisions. *Journal of Economic Behavior & Organization, 117*, 121–134.

Beal, D. J., Goyen, M., & Philips, P. (2005). Why do we invest ethically? *Journal of Investing, 14*, 66–78.

Becchetti, L., Ciciretti, R., Dalò, A., & Herzel, S. (2015). Socially responsible and conventional investments funds: Performance comparison and the global financial crisis. *Applied Economics, 47*, 2541–2562.

Bénabou, R., & Tirole, J. (2006). Incentives and prosocial behaviour. *The American Economic Review, 96*, 1652–1678.

Bénabou, R., & Tirole, J. (2011). Identity, morals, and taboos: Beliefs as assets. *The Quarterly Journal of Economics, 126*, 805–855.

Benson, K. L., & Humphrey, J. E. (2008). Socially responsible investment funds: Investor reaction to current and past returns. *Journal of Banking & Finance, 32*, 1850–1859.

Bollen, N. P. D. (2007). Mutual fund attributes and investor behaviour. *Journal of Financial and Quantitative Analysis, 42*, 683–708.

Bordalo, P., Gennaioli, N., & Shleifer, A. (2013). Salience and asset prices. *The American Economic Review, 103*, 623–628.

Borgers, A. C., & Pownall, R. A. (2014). Attitudes towards socially and environmentally responsible investment. *Journal of Behavioral and Experimental Finance, 1*, 27–44.

Boulstridge, E., & Carrigan, M. (2000). Do consumers really care about corporate responsibility? Highlighting the attitude-behaviour gap. *Journal of Communication Management, 4*, 355–368.

Camerer, C., Loewenstein, G., & Weber, M. (1989). The curse of knowledge in economic settings: An experimental analysis. *The Journal of Political Economy, 97*, 1232–1254.

Cheah, E. T., Jamali, D., Johnson, J. E. V., & Sung, M. C. (2011). Drivers of corporate social responsibility attitudes: The demography of socially responsible investors. *British Journal of Management, 22*, 305–323.

Cox, P., Brammer, S., & Millington, A. (2004). An empirical examination of institutional investor preferences for corporate social performance. *Journal of Business Ethics, 52*, 27–43.

Da, Z., Engelberg, J., & Gao, P. (2011). In search of attention. *The Journal of Finance, 66*, 1461–1499.

de Zwaan, L., Brimble, M., & Stewart, J. (2015). Member perceptions of ESG investing through superannuation. *Sustainability Accounting, Management and Policy Journal, 6*, 79–102.

Diouf, D., Hebb, T., & Touré, E. H. (2016). Exploring factors that influence social retail investors' decisions: Evidence from Desjardins fund. *Journal of Business Ethics, 134*, 45–67. https://doi.org/10.1007/s10551-014-2307-4

Duranda, R. B., Kohb, S. K., & Limkriangkraic, M. (2013). Saints versus sinners. Does morality matter? *Journal of International Financial Markets, Institutions and Money, 24*, 166–183.

Flammer, C. (2015). Does corporate social responsibility lead to superior financial performance? A regression discontinuity approach. *Management Science, 61*, 2549–2568.

Friedman, H. L., & Heinle, M. S. (2016). Taste, information, and asset prices: Implications for the valuation of CSR. *Review of Accounting Studies, 21*, 740–767.

Gil-Bazo, J., Ruiz-Verdú, P., & Santos, A. A. (2010). The performance of socially responsible mutual funds: The role offers and management companies. *Journal of Business Ethics, 94*, 243–263.

Glac, K. (2009). Understanding socially responsible investing: The effect of decision frames and trade-off options. *Journal of Business Ethics, 87*, 41–55.

Glac, K. (2012). The impact and source of mental frames in socially responsible investing. *Journal of Business Ethics, 13*, 184–198.

Golman, R., Hagmann, D., & Loewenstein, G. (2017). Information avoidance. *Journal of Economic Literature, 55*, 96–135.

GSIA. (2019). *2018 global sustainable investment review.* Retrieved from http://www.gsi-alliance.org/wp-content/uploads/2019/03/GSIR_Review2018.3.28.pdf Accessed 3 Nov 2020.

GSIA. (2020). *2020 global sustainable investment review.* Retrieved from http://www.gsi-alliance.org/wp-content/uploads/2021/08/GSIR-20201.pdf. Accessed 30 Apr 2022.

Heeb, F., Kölbel, J., Paetzold, F., & Zeisberger, S. (2021). *Do investors care about impact?* Retrieved from SSRN: https://ssrn.com/abstract=3765659. Accessed 24 Aug 2021.

Hirsch, D., & Terlau, W. (2015). Sustainable consumption and the attitude-behaviour-gap phenomenon—Causes and measurements towards a sustainable development. *International Journal on Food System Dynamics, 6*, 159–174.

Hirshleifer, D., & Teoh, S. H. (2003). imited attention, information disclosure, and financial reporting. *Journal of Accounting and Economics, 36*, 337–386.

Hofmann, E., Hoelzl, E., & Kirchler, E. (2008). A comparison of models describing the impact of moral decision making on investment decisions. *Journal of Business Ethics, 82*, 171–187.

Holm, C., & Rikhardsson, P. (2008). Experienced and novice investors: Does environmental information influence investment allocation decisions? *European Accounting Review, 17*, 537–557.

Hong, H., & Kacperczyk, M. (2009). The price of sin: The effects of social norms on markets. *Journal of Financial Economics, 93*, 15–36.

Jain, A., Jain, P. K., & Rezaee, Z. (2016). Value-relevance of corporate social responsibility: Evidence from short selling. *Journal of Management Accounting Research, 28*(2), 29–52. https://doi.org/10.2308/jmar-51439

Jansson, M., & Biel, A. (2014). Investment institutions' beliefs about and attitudes toward socially responsible investment (SRI): A comparison between SRI and Non-SRI Management. *Sustainable Development, 22*, 33–41.

Jones, R., & Murrell, A. J. (2001). Signalling positive corporate social performance. An event study of family-friendly firms. *Business &Society, 40*, 59–78.

Junkus, J. C., & Berry, T. C. (2010). The demographic profile of socially responsible investors. *Managerial Finance, 36*, 474–481.

Kahneman, D. (1973). *Attention and effort*. Prentice Hall.

Kahneman, D., & Tversky, A. (1979). Prospect theory: An analysis of decision under risk. *Econometrica, 47*, 263–291.

Keloharju, M., Knupfer, S., & Linnainmaa, J. T. (2012). Do investors buy what they know? Product market choices and investment decisions. *Review of Financial Studies, 25*, 2921–2958.

Kempf, A., & Osthoff, P. (2007). The effect of socially responsible investing on portfolio performance. *European Financial Management, 13*, 908–922.

King, B. G., & Pearce, N. A. (2010). The contentiousness of markets: Politics, social movements, and institutional change in markets. *Annual Review of Sociology, 36*, 249–267.

Kobler, D., Hauber, F., & Ernst, B. (2015). Millennials and wealth management: Trends and challenges of the new clientele. *Quarterly Insight Deloitte*. Retrieved from https://www2.deloitte.com/content/dam/Deloitte/lu/Documents/about-deloitte/Inside/lu-inside9-full.pdf. Accessed 3 Nov 2020.

LeBoeuf, R. A., & Shafir, E. (2003). Deep thoughts and shallow frames: On the susceptibility to framing effects. *Journal of Behavioral Decision Making, 16*, 77–92.

Lee, D. D., Humphrey, J. E., Benson, K. L., & Ahn, J. Y. (2010). Socially responsible investment fund performance: The impact of screening intensity. *Accounting and Finance, 50*, 351–370.

Lewis, A. (2001). A Focus Group Study of the Motivation to Invest: 'Ethical/Green and 'Ordinary' Investors Compared. *Journal of Socio-Economics, 30*, 331–341.

Linciano, N., Costa, D., Gentile, M., & Soccorso, P. (2019). *Report on financial investments of Italian household*. CONSOB Statistics and Analysis. Retrieved from http://www.consob.it/documents/46180/46181/rf2019.pdf/b3a1763b-a869-4aca-8930-9ae370a0aa90. Accessed 3 Nov 2020.

Martin, P. R., & Moser, D. V. (2016). Managers' green investment disclosures and investors' reaction. *Journal of Accounting and Economics, 61*, 239–254.

Michelson, G., Wailes, N., Van Der Laan, S., & Frost, G. (2004). Ethical investment processes and outcomes. *Journal of Business Ethics, 52*, 1–10.

McLachlan, J., & Gardner, J. (2004). A comparison of socially responsible and conventional investors. *Journal of Business Ethics, 52*, 11–25.

Morgan Stanley. (2017). *Millennials drive growth in sustainable investing*. Retrieved from https://www.morganstanley.com/ideas/sustainable-socially-responsible-investing-millennials-drive-growth/. Accessed 4 Nov 2020.

Morgan Stanley. (2019). *Sustainable reality, white paper*. Retrieved from https://www.morganstanley.com/content/dam/msdotcom/ideas/sustainable-investing-offers-financial-performance-lowered-risk/Sustainable_Reality_Analyzing_Risk_and_Returns_of_Sustainable_Funds.pdf. Accessed 3 Nov 2020.

Moskowitz, M. R. (1972). Choosing socially responsible stocks. *Business and Society Review, 72*, 71–75.

Neumann, B. R., Roberts, M. L., & Cauvin, E. (2011). Stakeholder value disclosures: Anchoring on primacy and importance of financial and nonfinancial performance measures. *Review of Managerial Science, 5*, 195–212.

Nilsson, J. (2008). Investment with a conscience: Examining the impact of pro-social attitudes and perceived financial performance on socially responsible investment behaviour. *Journal of Business Ethics, 83*, 307–325.

Nilsson, J. (2009). Segmenting socially responsible mutual fund investors: The influence of financial return and social responsibility. *International Journal of Bank Marketing, 27*, 5–31.

Pasewark, W. R., & Riley, M. E. (2010). It's a matter of principle: The role of personal values in investment decisions. *Journal of Business Ethics, 93*, 237–253.

Peifer, J. L. (2014). Fund loyalty among socially responsible investors: The importance of the economic and ethical domains. *Journal of Business Ethics, 121*, 635–649.

Petersen, H. L., & Vredenburg, H. (2009). Morals or economics? Institutional investor preferences for corporate social responsibility. *Journal of Business Ethics, 90*, 1–14.

Pilaj, H. (2017). The choice architecture of sustainable and responsible investment: Nudging investors toward ethical decision-making. *Journal of Business Ethics, 140*, 743–753.

Renneboog, L., Horst, J. T., & Zhang, C. (2008). The price of ethics and stakeholder governance: The performance of socially responsible mutual funds. *Journal of Corporate Finance, 14*, 302–322.

Renneboog, L., Horst, J. T., & Zhang, C. (2011). Is ethical money financially smart? Nonfinancial attributes and money flows of socially responsible investment funds. *Journal of Financial Intermediation, 20*, 562–588.

Richardson, B. (2007). Do the fiduciary duties of pension funds hinder socially responsible investment? *Banking and Finance Law Review, 22*, 146–201.

Richardson, B. (2009). Keeping ethical investment ethical: Regulatory issues for investing for sustainability. *Journal of Business Ethics, 87*, 555–572.

Richardson, B. (2011). From fiduciary duties to fiduciary relationships for socially responsible investing: Responding to the will of beneficiaries. *Journal of Sustainable Finance and Investment, 1*, 5–19.

Riedl, A., & Smeets, P. (2017). Why do investors hold socially responsible mutual funds? *The Journal of Finance, 72*, 2505–2550.

Rosen, B. N., Sandler, D. M., & Shani, D. (1991). Social issues and socially responsible investment behaviour: Preliminary empirical investigation. *Journal of Consumer Affairs, 25*, 221–234.

Rubaltelli, E., Lotto, L., Ritov, I., & Rumiati, R. (2015). Moral investing: Psychological motivations and implications. *Judgment and Decision Making, 10*, 64–75.

Rzeźnik, A., Hanley, K., & Pelizzon, L. (2021). *The salience of ESG ratings for stock pricing: Evidence from (potentially) confused investors*. SAFE Working Paper No. 310. Retrieved from https://ssrn.com/abstract=3801703. Accessed 26 Aug 2021.

Säve-Söderbergh, J. (2010). Who lets ethics guide his economic decision-making? An empirical analysis of individual investments in ethical funds. Economics Letters, 107, 270–272.

Schelling, T. C. (1960). *The strategy of conflict*. Harvard University Press.

Schueth, S. (2003). Socially responsible investing in the United States. *Journal of Business Ethics, 43*, 189–194.

Simon, H. A. (1955). A behavioral model of rational choice. *Quarterly Journal of Economics, 69*, 99–118.

Sparkes, R., & Cowton, C. J. (2004). The maturing of socially responsible investment: A review of the developing link with corporate social responsibility. *Journal of Business Ethics, 52*, 45–57. https://doi.org/10.1023/B:BUSI.0000033106.43260.99

Statman, M. (2008). *The expressive nature of socially responsible investors*. Working Paper, Santa Clara University. Retrieved from https://papers.ssrn.com/sol3/papers.cfm?abstract_id=1094068. Accessed 4 Nov 2020.

Tahir, I., & Brimble, M. (2011). Islamic investment behaviour and the permissibility of equity investments. *International Journal of Islamic and Middle Eastern Finance and Management, 4*, 116–130.

Tippet, J., & Leung, P. (2001). Defining ethical investment and its demography in Australia. *Australian Accounting Review, 11*, 44–55.

Tversky, A., & Kahneman, D. (1992). Advances in prospect theory: Cumulative representation of uncertainty. *Journal of Risk and Uncertainty, 5*, 297–323.

Wang, M., Qiu, C., & Kong, D. (2011). Corporate social responsibility, investor behaviors, and stock market returns: Evidence from a natural experiment in China. *Journal of Business Ethics, 101*, 127–141.

Webley, P., Lewis, A., & Mackenzie, C. (2001). Commitment among ethical investors: An experimental approach. *Journal of Economic Psychology, 22*, 27–42.

Williams, G. (2007). Some determinants of the socially responsible investment decision: A cross-country study. *The Journal of Behavioral Finance, 8*, 43–57.

Woodward, T. (2000). *The profile of individual ethical investors and their choice of investment criteria*. Bournemouth University.

Young, W., Hwang, K., McDonald, S., & Oates, C. J. (2010). Sustainable consumption: Green consumer behaviour when purchasing products. *Sustainable Development, 18*, 20–31.

PART III

Current Developments in the European Regulatory Framework

CHAPTER 8

Financial Regulation for Sustainable Finance in the European Landscape

Lucia Alessi, Barbara Alemanni, and Giovanna Frati

Introduction

In the face of the unprecedented challenge posed by the COVID-19 pandemic, the European Green Deal, together with digitalization, is a pillar of Europe's Recovery Plan. Given the central role of the European Green Deal in Europe's recovery strategy, financing it is a key policy priority for the Commission. Indeed, it is estimated that an additional

Barbara Alemanni authored section "The Sustainable Finance Disclosure Regulation (SFDR)"
Giovanna Frati authored section "The Delegated Legislation Integrating Sustainability into UCITS Directive and AIFMD"

L. Alessi (✉)
European Commission, Bruxelles, Belgium
e-mail: lucia.alessi@ec.europa.eu

N. Linciano et al. (eds.), *Information as a Driver of Sustainable Finance*, Palgrave Studies in Impact Finance,
https://doi.org/10.1007/978-3-030-93768-3_8

€470 billion will be needed yearly to meet the EU's climate and environmental goals.[1] In particular, to reach the current 2030 climate and energy targets—notably, a reduction in greenhouse gas (GHG) emissions of 55% compared to 1990, a share of at least 32% of renewable energy consumption and at least 32.5% energy savings compared to the business-as-usual scenario—€240 billion additional annual investment will be needed.[2] To this, an additional €100 billion per year need to be added to deliver on Europe's wider transport infrastructure and €130 billion per year to deliver on wider environmental objectives including environmental protection, resource management and circular economy. Given the magnitude of the challenge, public money alone will clearly not be enough to finance the green transition.

Hence, following the 2018 Action Plan on Financing sustainable growth, a Renewed Sustainable Finance Strategy has been designed to provide a roadmap with new actions, to increase private investment in sustainable projects and activities. This strategy can only be successfully implemented with the involvement of all stakeholders, including financial market participants, and based on sound scientific evidence.

Against this backdrop, this chapter reviews recent key regulatory developments in the European policy landscape in the area of sustainable finance. It starts by outlining the European Action Plan for financing sustainable growth, describing its rationale and the implementing measures taken so far. It then focusses on the EU Taxonomy for sustainable activities, which is the cornerstone of the Action Plan. With respect to corporate sustainability-related information, it provides a summary of relevant EU legislation, affecting both non-financial and financial corporations. Then, it offers an outlook on the Renewed Sustainable Finance Strategy, which will take existing actions forward and provide the needed policy tools to ensure a green recovery from the impact of the

B. Alemanni
Bocconi University, Milan, Italy
e-mail: barbara.alemanni@unibocconi.it

University of Genoa, Genoa, Italy

G. Frati
CONSOB, Rome, Italy
e-mail: g.frati@consob.it

COVID-19 pandemic. It concludes with some reflections on how scientific research can support the development of science-based legislation in the area of sustainable finance and help policymakers to overcome the main challenges.

The 2018 Sustainable Finance Action Plan

In December 2016, a year after the signature of the Paris Agreement, the European Commission established a High-level expert group on sustainable finance (HLEG). In January 2018, the HLEG delivered its advice on how the Commission could '*steer the flow of capital towards sustainable investments; identify steps that financial institutions and supervisors should take to protect the financial system from sustainability risks; and deploy these policies on a pan-European scale*'.[3]

Based on the HLEG recommendations, in March of the same year, the Commission adopted an Action Plan on financing sustainable growth, which sets out a comprehensive strategy to integrate finance and sustainability.[4] The Action Plan is structured around three objectives, and ten actions overall are associated with these objectives. Table 8.1 offers an overview of the Action Plan, including the three objectives, the associated actions and the corresponding legislative initiatives.

Reorienting Capital Flows Towards a More Sustainable Economy

The first objective of the Action Plan is to reorient capital flows towards sustainable investment. The key action foreseen in the Action Plan to achieve this objective is the establishment of a classification system for sustainable activities, the so-called 'Taxonomy', which defines what is considered sustainable—noting that the concept of sustainability does not only include the environmental dimension but also the social and the governance dimensions. So far, the Taxonomy has been developed for the two climate-related objectives of climate change mitigation and adaptation, and work is ongoing to extend it to other environmental objectives (see section "Sustainability-Related Disclosures" for details).

Still under the first objective, to protect the integrity of and trust in the sustainable financial market, as well as enable easier access for investors seeking sustainable financial products, Action 2 consists of the creation of standards and labels. In particular, it mentions the creation of an EU Green Bond Standard and an EU Ecolabel for financial products.

Table 8.1 Overview of the 2018 Action Plan on Sustainable Finance

Objective	*Action*	*Legislation*
Reorienting capital flows towards sustainable investment	EU classification system for environmentally sustainable economic activities	Taxonomy Regulation (adopted) and Delegated Regulation
	Standards and labels	EU Green Bond Standard (proposal adopted) Ecolabel for financial products (Decision to be adopted)
	Fostering investment in sustainable projects	Sustainable Europe Investment Plan (adopted)
	Incorporating sustainability in financial advice	Delegated Acts amending MiFID II and IDD to ensure that sustainability preferences are taken into account in the suitability assessment (adopted)
	Sustainability benchmarks	Benchmark Regulation introducing EU Climate Transition Benchmark and EU Paris-aligned Benchmark (adopted)
Mainstreaming sustainability into risk management	Sustainability in research and ratings	ESMA updated guidelines improving transparency on the role of sustainability in credit rating actions
	Disclosures by financial market participants	Sustainable Finance Disclosure Regulation (adopted) and Regulatory Technical Standards
	Sustainability in prudential requirements	Ongoing
Fostering transparency and long-termism	Strengthening sustainability disclosures by corporates	Corporate Sustainability Reporting Directive (proposal adopted)
	Fostering sustainable corporate governance and promoting long-termism	Corporate sustainability due diligence (proposal adopted)

The Commission has adopted a proposal for a Regulation on an EU Green Bond Standard[5] in July 2021, based on the recommendations of the Technical Expert Group (TEG) on Sustainable Finance.[6] Based on this proposal, an EU Green Bond could be issued by European and non-European issuers, as long as it complies with the following four requirements. The first requirement concerns the use of proceeds, which should be exclusively used to finance (or refinance) projects or activities defined as 'green' by the EU Taxonomy. The second requirement consists of a detailed reporting, providing details on all relevant aspects linked to the issuance, including the use of proceeds. The third requirement concerns mandatory third-party verification of the green bond, to ensure compliance with the Regulation. The fourth requirement relates to external reviewers, which must be registered with and supervised by the European Securities and Markets Authority (ESMA).

The EU Ecolabel for Financial products is being developed by the Commission. Like all EU Ecolabels, the Ecolabel for financial products aims to reward the top environmentally performing products in the relevant market. Based on the most recent Technical Report, published in March 2021,[7] the financial products in scope will be the following: (i) packaged retail and insurance-based investment products (PRIIPs), including investment funds[8] and insurance-based investment products and (ii) fixed-term or savings deposits. For a financial product to be eligible for the EU Ecolabel, it will need to comply with a set of criteria, which cover the following: a minimum share of financing going to 'green' activities, as defined by the Taxonomy; exclusions based on environmental, social and governance aspects; engagement and investor impact and mandatory information. The extension of the Ecolabel framework to financial products, by way of a Commission Decision, is expected for 2022.

Action 3 consists of fostering investment in sustainable projects. To this aim, the Commission has launched the Sustainable Europe Investment Plan, which is the investment pillar of the Green Deal (see Chapter 2). The Plan will mobilize at least €1 trillion in sustainable investments over the next decade and part of it—the Just Transition Mechanism—will be targeted to a fair and just green transition. InvestEU is a related instrument which will provide an EU budget guarantee to allow the European Investment Bank and other implementing partners to invest in more and higher-risk projects. The ultimate objective is to crowd-in private investors, by providing not only funding but also technical support and assistance.

Action 4 relates to incorporating sustainability in financial advice. To clarify the duties of financial advisors to provide their clients with clear advice on the social and environmental risks and opportunities attached to their investments, and to ensure that financial firms include sustainability in their procedures, the Commission adopted rules integrating sustainable finance into the Markets in Financial Instruments Directive (MiFID II) and the Insurance Distribution Directive (IDD) frameworks.[9] The amendments include provisions integrating sustainability in the suitability test, i.e. the assessment of whether particular financial services, instruments or products are 'suitable' for a particular client. They also integrate the assessment of sustainability risks and factors into the organizational requirements, and risk management and product governance procedures of investment firms.

Action 5, the last one under the first objective of the Action Plan, is to develop sustainability benchmarks. In this respect, two types of low-carbon benchmarks have been created, namely the EU Climate Transition Benchmark and the EU Paris-aligned Benchmark.[10] These are two voluntary labels that can be used for benchmarks that comply with a set of conditions. In the first case, the underlying assets need to be selected in a way that the resulting portfolio is on a decarbonization trajectory. In the latter case, the underlying assets need to be selected in a way that the resulting portfolio's carbon emissions are aligned with the objectives of the Paris Agreement. As such, the EU Paris-aligned benchmark is more ambitious than the EU Climate transition benchmark. Indeed, while the EU Climate Transition Benchmark requires a 30% lower GHG intensity than the investable universe, this is raised to 50% for the EU Paris-aligned Benchmark. This latter also needs to exclude companies deriving a significant share of their revenues from fossil fuels. As additional requirements, both benchmarks must use the 1.5 °C reference scenario and target a yearly reduction in portfolio's GHG emissions of at least 7%.[11]

Mainstreaming Sustainability into Risk Management

The second objective of the Action Plan is mainstreaming sustainability into risk management, i.e. including environmental and social considerations in decision-making processes with the aim to limit the financial impact of risks related to the sustainability dimension. Actions 6 to 8 of the Action Plan fall under this objective.

Action 6 deals with better integrating sustainability in ratings and market research. To this aim, the ESMA published updated guidelines to improve the transparency of credit rating actions concerning the extent to which sustainability factors have been taken into account.[12] With respect to ESG ratings, the European Commission published a study emphasizing the overall demand for greater transparency on the methodologies adopted by the providers of sustainability-related products, and their conflicts of interest, as well as for the development of industry standards.[13] More recently, ESMA published a letter to the European Commission calling for legislative action on ESG ratings and assessment tools to address risks of greenwashing, capital misallocation and products mis-selling.[14]

Action 7 is to clarify asset managers' and institutional investors' duties regarding sustainability. The Regulation on sustainability-related disclosures in the financial services sector, also known as Sustainable Finance Disclosure Regulation (SFDR), addresses this point (see section"Sustainability-Related Disclosures" for details).

Action 8 focusses on linking capital requirements for banks and insurers with sustainability considerations. In particular, the European Banking Authority (EBA) was asked to assess the potential inclusion of ESG risks in the supervisory review and evaluation process and to explore the prudential soundness of introducing a more risk-sensitive treatment of 'green assets' (so-called green supporting factor). In the context of the implementation of its Action Plan,[15] EBA published discussion paper on ESG risk management and supervision[16] and a consultation paper on prudential disclosures,[17] while the European Insurance and Occupational Pensions Authority (EIOPA) published an opinion on sustainability in the context of EU insurance regulation.[18]

Fostering Transparency and Long-Termism

The third objective of the Action Plan is fostering transparency and long-termism, i.e. the focus on longer-term objectives. In order to enable investors to properly assess the long-term value creation of companies, increased transparency on firm's management of sustainability risks—which are often longer-term—is essential.

To this aim, Action 9 in the Action Plan is to strengthen sustainability disclosure and accounting rule-making. The European Commission has recently adopted a new Corporate Sustainability Reporting Directive, which in practice replaces the Non-Financial Reporting Directive and will deliver more and better information from companies about their social and environmental performance and impacts (see section "The Delegated Legislation Integrating Sustainability into UCITS Directive and AIFMD" and Chapter 9 for details).

Finally, Action 10 is to foster sustainable corporate governance and attenuate short-termism in capital markets. On this issue the ESAs published their advices, emphasizing the importance of ESG disclosures to facilitate institutional investors' engagement.[19] The European Commission has adopted a proposal for a Directive on Corporate sustainability due diligence, which would provide incentives to company boards to properly integrate sustainability risks and opportunities into their strategy, with a view to extending the horizon of corporate decision-making.[20]

The EU Taxonomy for Sustainable Activities—The Need for a Common Definition

The EU Taxonomy is a unified classification system for sustainable activities, providing a definition of what can be considered 'sustainable'. It is a tool, just like a dictionary, where all economic agents, from firms to investors, can look up whether a particular economic activity can be defined sustainable, and under which conditions. The Taxonomy is the cornerstone of the Action Plan, as it is instrumental to the implementation of other actions, which refer to the Taxonomy for the definition of sustainable activities, and necessary for sustainable finance to reach scale.

The TEG on Sustainable Finance was established by the European Commission in 2018 and tasked with the development of a Taxonomy proposal, which was published in 2020. The Taxonomy Regulation[21] entered into force in July 2020, establishing the high-level criteria for determining whether an economic activity qualifies as environmentally sustainable, entered into force.

Structure and Features of the EU Taxonomy

The environmental Taxonomy is structured around the following six objectives:

- climate change mitigation;
- climate change adaptation;
- sustainable use and protection of water and marine resources;
- transition to a circular economy;
- pollution prevention and control;
- protection and restoration of biodiversity and ecosystems.

An activity is considered environmentally sustainable if it makes a substantial contribution to at least one of the six environmental objectives while not causing significant harm to any of the other five objectives. In order for these two conditions to be satisfied, an economic activity must fulfil a set of activity-specific technical screening criteria. Additionally, it needs to meet a set of minimum social safeguards, namely the OECD Guidelines for Multinational Enterprises and the UN Guiding Principles on Business and Human Rights.

The Taxonomy recognizes the following three different types of activities as environmentally sustainable:

- Activities that avoid or reduce emissions, such as zero-emission transport; or that increase GHG sequestration, such as carbon capture and storage technologies, or that use sustainably sourced renewable materials;
- Activities that enable other activities to make a substantial contribution to one or more objectives, for example, the sorting of non-hazardous waste, which enables the actual material recovery from the already sorted non-hazardous waste;
- Transitional activities, i.e. activities for which there is no technologically and economically feasible low-carbon alternative, but that have the minimum level of emissions compared to their sector. An example could be the manufacturing of iron and steel by means of the least emitting technology available.

The assessment of the contribution of an economic activity to a particular environmental objective needs to be based on lifecycle considerations.

Moreover, in order to be considered environmentally sustainable, a transitional activity shall: (i) not lead to a lock-in in carbon-intensive assets, also taking into account that technical screening criteria for transition activities will be progressively tightened; (ii) have GHG emission levels that correspond to the best performance in the sector or industry; (iii) not hamper the development and deployment of low-carbon alternatives; (iv) not have a technologically and economically feasible alternative. Finally, power generation activities from solid fossil fuels are explicitly excluded from being Taxonomy-eligible.

The TEG Taxonomy proposal focussed on the two climate-related objectives of climate change mitigation and adaptation. The TEG decided to structure the Taxonomy broadly following the NACE classification, i.e. the statistical classification of economic activities in the European Community. However, the definition of relevant economic activities in the Taxonomy does not always map one-to-one to the NACE classification.

The TEG proposal was the basis for the first Delegated Act under the Taxonomy Regulation, so-called Climate Delegated Act, containing technical screening criteria for a first list of activities making a substantial contribution to climate change mitigation and adaptation.[22] In particular, the Climate Delegated Act covers economic activities which are responsible for almost 80% of direct GHG emissions in Europe. The technical screening criteria for the 'substantial contribution' and the 'do no significant harm' conditions are often stricter than the provisions of sectoral legislation, resulting in a very ambitious Taxonomy. This is because a science-based environmental sustainability assessment, with a long-term horizon, shows that often more stringent criteria are needed to reach the EU environmental goals.[23]

At the same time, the Taxonomy expands the green investable universe much beyond renewable energy, covering a broad range of economic activities in various sectors, including manufacturing, transport, buildings and many others. This is a key characteristic of the Taxonomy, as it allows environmentally minded investors to diversify their portfolios. For example, the taxonomy covers also a set of highly emitting activities, such as cement production, specifying that these activities can also be 'green' if they are carried out by minimizing process emissions through energy efficiency improvements and switch to alternative fuels.

Some activities were left out of the Climate Delegated Act as more discussion was needed. In particular, technical screening criteria for Agriculture, which were included in the draft Delegated Act put forward for consultation, have eventually been put on hold pending further progress on the negotiations underway on the Common Agricultural Policy. With respect to nuclear energy, the Joint Research Centre (JRC) of the European Commission was tasked with an assessment of nuclear energy from a 'do no significant harm' perspective considering the entire nuclear life cycle, and concluded that there is no 'science-based evidence that nuclear energy does more harm to human health or to the environment than other electricity production technologies already included in the Taxonomy'.[24] The JRC report has been reviewed by two expert groups, which have published their opinions in July 2021.[25] These three inputs have informed the Commission adoption of a so-called Complementary Delegated Act including, under strict conditions, specific nuclear energy activities in the list of economic activities covered by the EU Taxonomy.[26] The Complementary Delegated Act also includes certain gas activities, under the condition that they contribute to the transition from coal to renewables, in line with the European Council's conclusion that gas is a transitional technology.[27]

The Sustainable Finance Platform, an expert group set up by the European Commission to replace the TEG, is currently working on developing the Taxonomy for the other four environmental objectives, building on a framework developed by the JRC.[28] Another Delegated Act will then be adopted containing the list of relevant activities for these additional objectives, as well as possible activities providing a substantial contribution to the climate objectives which were not included in the first Delegated Act (see section "Final Remarks Concerning ESG Corporate Disclosures").[29] More broadly, the Taxonomy is a living tool and its criteria and scope will be subject to regular review to reflect market and technological developments.

Uses of the EU Taxonomy

The Taxonomy Regulation lists some mandatory uses of the Taxonomy, which relate to sustainability disclosures for green financial products and larger companies (see next section as well as Chapters 9and10). At the same time, it should be emphasized that the use of the Taxonomy by financial market participants as a tool to orient their investment strategies remains purely voluntary. The Taxonomy is not intended to be a mandatory list to invest in and in fact, it can support different investment styles and strategies. Also, it should be stressed that the Taxonomy does not make any judgement on the financial performance of an investment, but only on its environmental performance. Finally, the Taxonomy only classifies activities, not companies and any company can carry out Taxonomy-aligned activities (e.g. purchasing an electric-vehicle fleet or installing roof solar panels).

Having said that, investors are encouraged to use the Taxonomy to select their holdings and to engage with investee companies to support their green transition, as the Taxonomy does provide a means to measure the greenness of an investor exposures. It should be stressed that, while the ultimate objective of the sustainable finance agenda is to increase financing to sustainable activities, and while some activities should be progressively abandoned in the transition to a zero-carbon economy, this does not necessarily imply divesting from particular companies. On the contrary, companies that need to transition will in fact benefit from the existence of the Taxonomy, as those capital expenses that these companies will bear to 'green' their business may be counted as 'green capital expenditure (CapEx)' and hence, the financing of these activities might fully count as green finance.

Moreover, financial market participants can use the Taxonomy to design green financial products, such as green bonds, green loans or energy-efficient mortgages, which will allow any issuer/borrower to improve its environmental performance. Finally, investors can assess the greenness of their exposures by using the Taxonomy, as it allows to make a link between a financial asset, be it a loan or a security, and its environmental performance. For project-loans and green bonds, the Taxonomy-alignment of the financial instrument corresponds to the alignment of the project/activity that the product ultimately finances. For general purpose loans, bonds and equity exposures, the Taxonomy-alignment of the issuer/borrower (primarily based on its turnover, alternatively on its CapEx) is taken as a proxy for the Taxonomy-alignment of the financial asset.

However, this assessment is far from straightforward. Indeed, another stream of the Sustainable Finance Platform is working on addressing issues related to the practical use of the taxonomy and to Taxonomy-based reporting. For example, companies often use other industrial classifications than NACE, such as BICS and GICS,[30] and yet other classifications may be used by data providers. Another challenge consists of identifying suitable enabling activities when using a whole value chain approach.

But the probably biggest challenge at this stage stems from the absence of granular data on the Taxonomy-alignment of investee companies. Indeed, while relevant data will become increasingly available with the implementation of EU regulation on sustainability-related disclosures over the coming years (see next section as well as Chapter 9), at the moment virtually no data exist on the Taxonomy-alignment of any company. In this context, large financial corporations will be required to disclose their own taxonomy-alignment, based on the Taxonomy-alignment of their counterparts, on which they have little or no information. A pragmatic solution could be the use of proxies and estimates, which are however only available, at best, for larger firms, and based on proprietary models which are not always fully transparent.

Finally, although the Taxonomy was designed as a tool to crowd-in private investment into sustainable activities, it is becoming apparent that its usefulness may go beyond the private sector. No obligation exists for Governments to disclose based on the Taxonomy, public finances are not classified based on NACE sectors, and relevant information to assess the Taxonomy-alignment of publicly-financed projects and activities does not exist. Still, the Taxonomy could be taken as a benchmark in sustainability-proofing and assessments of investment plans and projects. In particular, the EU Recovery and Resilience Facility explicitly refers to the EU Taxonomy, and in particular to the principle of 'do no significant harm', to ensure that investments are in line with the long-term ambitions of the European Green Deal (see Chapter 2).

From the Taxonomy to the Taxonomies?

It should be noted that, while the Taxonomy identifies 'green' activities and further classifies them into low-carbon, enabling and transitional activities, it says nothing about the environmental impact of the activities it does not classify. In other words, activities that are not aligned with the Taxonomy criteria are simply 'not green'. However, those activities comprise different types, namely:

- Activities that are covered by the Taxonomy but do not meet the technical screening criteria, whether those for 'substantial contribution', those on 'do no significant harm', or both. For example, the manufacturing of electric vehicles is not aligned with the Taxonomy if it uses toxic substances and generates toxic waste ('do no significant harm' assessment). An example of failure in meeting the 'substantial contribution' criteria for climate change mitigation could be an aluminium manufacturing process characterized by higher emissions than the relevant threshold;

- Activities that are not in the scope of the Taxonomy because they have a low environmental impact, such as education;
- Activities that are not in the scope of the Taxonomy simply because the Taxonomy is not yet complete, but could well be included in the future with the development of relevant technical screening criteria;
- Activities that are harmful with respect to one or more objectives, such as activities that pollute air, water or soil.

The Sustainable Finance Platform is currently discussing whether it could be useful to extend the Taxonomy framework to classify also non-green activities and if yes, how one could go about it (see section "The Delegated Legislation Integrating Sustainability into UCITS Directive and AIFMD").[31] In some cases, the technical screening criteria on the 'do no significant harm' condition could serve as a reference when determining the thresholds for a 'significant harm' space. Among the stakeholders who could benefit from the development of a 'harmful' taxonomy there are, for example, those non-financial corporates currently carrying out harmful activities: should 'harmful' be defined by means of technical screening criteria, these companies could use these thresholds to benchmark and describe their environmental performance, showing that they are transitioning too. Financial supervisors and financial institutions themselves could also find a 'harmful' taxonomy useful, as it would allow them to better identify riskier exposures. Indeed, by covering the climate change mitigation and adaptation objectives, a 'harmful' taxonomy would help assessing the so-called 'transition' and 'physical' risks. As detailed in Chapter 3, the former are those financial risks that could materialize in the medium-term, should more decisive green policies lead to higher costs for more polluting firms, causing a generalized worsening in the financial performance of these firms. In parallel, carbon-intensive assets would increasingly become 'stranded'. The latter are those financial risks that stem from more frequent and severe climate-related natural catastrophes (e.g. typhoons and floods), which could negatively affect the performance of assets linked to particularly vulnerable economic activities or geographical areas.

Finally, the Sustainable Finance Platform is also working on a proposal for a social Taxonomy, including the governance dimension. Its objectives could include education, health, housing, equitable and safe employment, and others (see section "The Strategy for Financing the Transition to a Sustainable Economy").[32]

Sustainability-Related Disclosures

An important element shared across all the three, interrelated, objectives of the Action Plan, is transparency and availability of sustainability-related information. More comprehensive information on ESG, including climate-related issues, will lead to greater transparency and benefit financial market participants in various ways. Non-financial corporates and issuers more generally will be able to mitigate reputational risks, by disclosing the extent to which their business is aligned with the Taxonomy and on their plans to improve their environmental performance. By having to disclose these points, companies will need to start a reflection on ESG risks and how they might impact their business going forward. A better awareness of their overall risk exposure should arguably lead to an ultimately improved performance, in an economic environment that is transitioning to a sustainable development paradigm. Clearer and more comprehensive information will also contribute to reduce 'greenwashing' in the market as a whole, rewarding those companies that are really engaged in the green transition.

On the side of investors, increased availability of information on the environmental performance of investee companies will lead to more accurate credit risk assessments and will make them save the time and money that should have been otherwise spent to properly incorporate ESG risks in investment strategies. An easier access to relevant information will enable investors to make more informed decisions by understanding a company's development, performance, position and impact. This will ultimately help them to avoid the underestimation of the long-term effects that sustainability factors might have on the financial performance of their investments.

Given the central role of sustainability-related information, several legislative actions have been taken to codify the implementation of the Action Plan's objectives at the EU level. In particular, there are five pieces of legislation that affect ESG disclosures for financial market participants, as well as larger financial and non-financial corporates.

The New Corporate Sustainability Reporting Directive (CSRD)

The first piece of EU regulation introducing sustainability reporting rules for larger companies was the Non-Financial Reporting Directive (NFRD; more on this in Chapter 9).[33] It established a key principle, i.e. the 'double materiality perspective', meaning that companies must disclose about how environmental and social risks affect their business ('outside-in') and about their own impact on people and the environment ('inside-out'). In order to assist companies to disclose non-financial information in a relevant, useful, consistent and more comparable manner, the European Commission published non-binding guidelines in 2017,[34] followed in 2019[35] by a supplement on reporting climate-related information.

However, given the attention that sustainability has gained in recent years and the increased importance that investors attach to sustainability-related information, the NFRD framework had become insufficient. Hence a proposal for a new Corporate Sustainability Reporting Directive (CSRD)[36] was adopted, which replaces the NFRD and strengthens EU legal provisions on the disclosure of sustainability-related information by companies. In particular, the CSRD proposal extends reporting requirements in the following dimensions:

- Scope: while the NFRD only applied to larger companies, defined as those with more than 500 employees (about 11,000 companies), the CSRD scope includes all large companies (also unlisted) and listed companies. This includes listed SMEs, but with the exception of listed micro-companies, as well as companies not established in the EU that are listed on EU-regulated markets, and the EU subsidiaries of non-EU companies, for a total of around 49,000 firms.
- Auditing: companies in the CSRD scope will need to seek limited assurance for reported information, meaning that the procedures the auditor will perform will not address the details and be narrower in scope than those performed in a reasonable assurance engagement. However, the CSRD includes an option to move towards a reasonable assurance requirement at a later stage.
- Standards: companies in the CSRD scope will need to report in accordance with EU standards, which will be simpler for SMEs. EU

sustainability reporting standards will be developed by the European Financial Reporting Advisory Group (EFRAG), and will be tailored to EU policies, while building on and contributing to international standardisation initiatives.[37] Moreover, information will need to be made available in a digital format (machine readable).

A first set of standards is planned for adoption by 31 October 2022, while a second set, together with standards for SMEs, should be adopted one year later, with the first company reports expected by 2024 covering the financial year 2023.

The Sustainable Finance Disclosure Regulation (SFDR)

To create a common language around sustainability for financial institutions the regulatory reform put forward a 'Disclosure Regulation' (SFDR), which has applied since March 2021.[38] Additionally in April 2021, within an ambitious and comprehensive package of measures to help improve the flow of money towards sustainable activities across the European Union, the Commission also adopted six amending Delegated Acts on fiduciary duties, investment and insurance advice that will ensure that financial firms, e.g. advisers, asset managers or insurers, include sustainability in their procedures and their investment advice to clients.[39] In February 2021 the ESAs published a Report on draft regulatory technical standards (RTS) under the SFDR and proposed that the RTS should apply from 1 January 2022.[40] On 22 October 2021 the Joint Committee of the ESAs published the Regulatory Technical Standards under the SFDR and the Taxonomy Regulation, concerning the content and presentation of information on products that promote or contribute to environmental objectives through sustainable investments. The taxonomy of eco-friendly activities should be complemented by a social taxonomy and a taxonomy of environmental transition. A second delegated act on the technical screening criteria for the remaining environmental objectives will be published in the course of 2022.

The SFDR seeks to harmonize existing provisions on disclosures to investors in relation to sustainability-related disclosures by imposing requirements on so-called financial market participants (e.g. AIFMs and UCITS management companies and investment firms carrying out portfolio management) and financial advisers (firms authorized under MiFID to give investment advice and credit institutions) in relation to financial products (e.g. AIFs, UCITS). The SFDR does not require any changes to the way products are managed, however it requires an enhanced level of disclosure. The nature and extent of these disclosure requirements depend on product classification under the regulation.

Products that have met certain ESG criterion with respect to investment process or objective may be classified as Article 8 ('light green' products) or Article 9 products ('dark green' products). All products which do not meet these criteria are classified as Article 6 or 'other' products. Article 8 products consider or promote environmental or social characteristics in the pursuit of other financial objectives. Article 9 products seek to make a positive impact on society or the environment through sustainable investment and have a non-financial objective at the core of their offering. For both Article 8 and 9 products, ESG considerations are binding.

As it will be further elaborated in Chapter 10, the SFDR will require the integration of sustainability risks in financial market participants' investment decision-making processes or, where relevant, advisory processes and transparency as regards financial products which target sustainable investments, including a reduction in carbon emissions. Specific requirements include pre-contractual disclosures; disclosures on websites and disclosures in periodic reports in relation to financial products.

With this new transparency rules, a range of EU financial institutions will be required to post a number of sustainability-linked disclosures on their websites. These disclosures include:

- A policy on the firm's approach to sustainability risk.
- Data on whether, and if so how, the firm takes into account the 'principal adverse impacts' of its investment decisions or advice on sustainability.
- How the firm's remuneration policies are consistent with the integration of sustainability risks.

This new requirement will impact profoundly on how firms provide either advice or portfolio/asset management services since the required disclosures will need to summarize the integration of sustainability risks into the firm's investment decision-making processes, investment advice, or insurance advice, as relevant.

The EU goal is for ESG data to flow throughout the financial system; in other words, it should not simply be reserved for consumption by sophisticated professional investors, but should also be available in some form to retail end-investors. The European Commission has expressed concern that, at present, only a minority of clients receiving investment advice proactively raise ESG issues, and that there is currently a limited understanding among clients around the impact of ESG factors on risk and performance.[41] In the April 2021 Delegated Acts, two key methods that the EU is proposing to adopt in order to rectify this lack of engagement with end clients are:

1. Requiring firms to define a 'target market' for financial products by reference to ESG preferences.
2. Incorporating an assessment of each end-client's ESG preferences into the suitability test applying to investment advice, advice on insurance-based investment products and portfolio management services.

Requiring firms to engage with the expectations and preferences of end clients in this way could ultimately prove a powerful tool in shaping the focus given to ESG issues in the retail market. This will in turn form one of many drivers pushing corporate issuers towards a greater consideration of ESG, particularly given increasing industry moves to open up the equities market to retail investors.

With the reforms on product governance,[42] EU distributors will need to define target markets by reference to their ESG preferences 'where relevant'. In its technical advice, ESMA has left this test of relevance deliberately vague, however, noting that the amendments to the product governance regime 'are currently just a first step of a more extensive project,' and that this more flexible approach is 'meant as a starting point' that 'allows market participants to accommodate themselves to ESG-requirements in the context of Product Governance.'

On the suitability side, under the current regime, the information sought by firms about their clients' investment objectives will generally relate to financial objectives, while non-financial objectives (including ESG preferences) are rarely addressed. The EC reform, therefore, aims to build an assessment of each client's ESG preferences into the suitability test. Firms undertaking a suitability test will need to disclose information on the ESG characteristics of products offered to clients, and explain how the client's ESG preferences have been taken into account in selecting the product or providing the portfolio management service. Additionally, clients indicating sustainability preferences should be offered 'sustainable investment' products, which would mean article 8 or article 9 products.

With more opportunities to compare and contrast investment decisions according to their values, investors will be able to make more informed investment decisions. As they do, this may drive more capital into sustainable investments. This increases the growth of the sustainable economy. In addition, quality information can lead to improved financial performance and a better understanding of the impact of investments. Investors will have a much bigger window into the measures of their portfolio, analysis of risk and achievement of sustainable objectives over time.

The Taxonomy Regulation (TR)

The **Taxonomy Regulation** amends Articles 8, 9 and 11 of the SFDR requiring financial market participants to disclose more detailed, product-level information. In particular, for financial products that contribute to an environmental objective, information must be disclosed on which environmental objective(s) they contribute to and on how and to what extent these qualify as Taxonomy-aligned (Article 5). For financial products that promote environmental characteristics, in addition to the previous information, a statement shall be included for the part of the financial product which does not take into account the Taxonomy criteria for environmentally sustainable economic activities (Article 6). For all other financial products, it shall be stated in a specific, explicit manner that the Taxonomy is not taken into account (Article 7).

In addition, the Taxonomy Regulation obliges any undertaking which is subject to the CSRD to include in its non-financial statement or consolidated non-financial statement information on how and to what extent the undertaking's activities are associated with economic activities that qualify as environmentally sustainable (Article 8(1)). The Delegated Act which specifies the information that companies subject to the CSRD will have to disclose was adopted in July 2021[43] and the first investor disclosures using the EU Taxonomy are due at the beginning of 2022, covering the financial year 2021.

The Capital Requirements Regulation (CRR) and the Investment Firms Regulation (IFR)

The Capital Requirements Regulation (CRR)[44] obliges large institutions which have issued securities that are admitted to trading on a regulated market of any Member State to disclose information on ESG risks, including physical and transition risks (Article 449a). As per Article 434a, the EBA published Implementing Technical Standards (ITS) for Pillar 3 disclosures on ESG risks in February 2021. Starting from June 2022, the information shall be disclosed on an annual basis for the first year and twice a year from then on.

Similarly to the disclosure obligations of large firms which are subject to the CRR, the Investment Firms Regulation (IFR)[45] makes mandatory for investment firms to disclose information on ESG risks, including physical risks and transition risks. As of the 26 December 2022, the information shall be disclosed once in the first year and biannually thereafter.

The European Single Access Point (ESAP)

The wide range of information required by the aforementioned legislation to be disclosed by companies to the public has been, until now, disseminated in a fragmented manner and inconsistent disclosure formats, while the implementation of some parts of the regulation has been heterogeneous at national level. The resulting high search costs have a detrimental effect on both firms and investors and impede the mobilization of capital towards green and low-carbon investments. Aiming to facilitate the flow and use of information, the Commission has adopted a legislative proposal on the establishment of a European Single Access Point (ESAP) for companies' public financial and sustainability-related information.[46] The ESAP will address primarily information published by companies that are capital market participants, as well as SMEs looking for funding. It will encompass information disclosed pursuant to the CSRD and it could also be extended to other categories of ESG information disclosed by financial institutions.

The Delegated Legislation Integrating Sustainability into UCITS Directive and AIFMD

In April 2021, in the context of the Financial sustainable package,[47] the European Commission published the amendments to the delegated Acts of UCITS Directive and AIFMD in order to include ESG factors in the decision-making process of asset management companies.[48]

The kick-off of that amendments is the recognition of the existence of both sustainability risk and sustainability factors.[49] Accordingly, the legislator envisages that:

- the sustainability risks shall be integrated, on the basis of proportionality principle, in the management of the UCITS and AIFs;
- the senior management is responsible for that integration;
- the identification of conflicts of interests shall also include those types that may arise as a result of the integration of sustainability risks;
- the asset manager retains the necessary resources and expertise for the effective integration of sustainability risks;
- the due diligence embodies the consideration of sustainability risks as well as, where relevant, the principal adverse impacts of investment decisions on sustainability factors[50];
- the risk management policy considers the exposure of the funds to the sustainability risks.

Besides directives on collective investment schemes, it is worth noting that also the SFDR and its delegated Act[51] (Level 2 measures of Disclosure Regulation)—even if they are apparently devoted only to disclosure requirements—ends up influencing the substance and the structure of the investment process adopted by the asset manager.

Indeed, to the extent that the Level 2 measures of Disclosure Regulation require the market participants (in which definition asset managers are included) to disclose the criteria used to identify the adverse impacts and the methodology to their prioritization, it takes for granted that such criteria and methodologies are embodied in each step (definition, due diligence, implementation and monitoring) of decision-making process.

Likewise, the obligation to disclose (i) the strategy used to attain the ESG social characteristics/objective, (ii) the sustainability indicators used in view of the ex-ante selection of the investable financial instruments, (iii) the output of ex post measurement of the attainment of such ESG characteristics/objective, definitely entails that such strategies, indicators and output of measurements, before being disclosed to final investors, are fully included in the investment process. Accordingly, it could be concluded that the disclosure is just the outward representation of organizational measures and behavioural standards that the asset manager shall adopt.

The Strategy for Financing the Transition to a Sustainable Economy

The European Commission Strategy for financing the transition to a sustainable economy,[52] adopted in July 2021 (in the following, also Renewed Sustainable Finance Strategy), will be another key lever to mobilize financing for the Green Deal and shape a sustainable recovery. To this aim, it will provide the policy tools that the financial system needs to genuinely support the transition towards sustainability, in a context of recovery from the negative social and economic consequences of the pandemic. At the same time, it is reasonable to expect that policies developed in the context of the Sustainable Finance agenda will have an impact beyond the private sector (see Chapter 2).

Based on the consultation that was carried out in 2020 to inform the design of the strategy, there is a consensus among stakeholders that the COVID-19 recovery could indeed be an opportunity to redirect investments towards sustainability objectives by providing an enabling long-term policy framework and leveraging innovative technologies such as digitalization. At the same time, it is widely acknowledged that short-term profit-seeking practices and greenwashing are still quite widespread, and that the availability, comparability and quality of ESG data remain an issue.

The 2018 Action Plan is the starting point for the development of the Renewed Sustainable Finance Strategy, which predominantly focusses on the following four areas: (i) financing the transition of the real economy towards sustainability (Action 1), (ii) providing opportunities to individuals and SMEs to have greater access to sustainable finance (Action 2), (iii) improving the financial sector's resilience and contribution to sustainability (Actions 3, 4 and 5), and (iv) promoting an international consensus for an ambitious global sustainable finance agenda (Action 6).

Action 1 focusses on transition financing. The Commission has announced that it will consider proposing legislation to support the financing of certain economic activities, primarily in the energy sector, including natural gas, which contributes to reducing GHG emissions but cannot be considered sustainable.[53] In this respect, it should be noted that investments in the gas are eligible for funding support by EU instruments, such as those described in Chapter 2. Moreover, it will consider options to extend the Taxonomy to recognize environmental performance improvements, which however cannot be defined as 'green' as they fail to deliver a 'substantial contribution' (see section "Sustainability-Related Disclosures"). With respect to the further development of the Taxonomy, while a Complementary Climate Delegated Act has been adopted already to covers certain energy activities, a further Delegated Act will cover the four environmental objectives related to water, biodiversity, pollution prevention and circular economy (see section "The EU Taxonomy for Sustainable Activities—The Need for a Common Definition"). Finally, this action also includes the development of further standards and labels, including, e.g. transition or sustainability-linked bonds, as well as ESG benchmarks, and the definition of minimum sustainability criteria for financial products that promote environmental or social characteristics.

Action 2 aims at increasing the inclusiveness of the sustainable finance framework by leveraging the potential of citizens and SMEs. To this aim, the EBA will provide an opinion on the definition and possible supporting tools for green retail loans and green (energy efficient) mortgages. Moreover, the Commission will explore options to increase the access of citizens and SMEs to sustainable finance advisory services. Action 2 also proposes to use digital technologies to support SMEs and retail investors, e.g. by integrating ESG data in the data spaces under the European Data Strategy.[54] Action 2 also points at increasing insurance coverage against climate and environmental risks, by identifying potential insurance gaps through the EIOPA disaster insurance dashboard. With respect to the social dimension, a report will be published on the provisions required for a social taxonomy (see section "The EU Taxonomy for Sustainable Activities—The Need for a Common Definition"), while the Commission has already put forward a proposal on Corporate sustainability due diligence.[55] Finally, the Commission will strengthen tracking methodologies on climate and biodiversity in the context of the EU budget, encourage the use of green budgeting tools by the Member States and promote risk-sharing between public and private investors.

Action 3 relates to the resilience of the economic and financial system to sustainability risks. The proposed measures aim at reflecting sustainability risks in financial reporting standards and accounting, improving transparency of credit ratings and rating outlooks, enhancing banks' and insurers' identification and management of sustainability risks, and managing ESG risks at system level. To do so, the Commission will work with EFRAG, ESMA and IASB on capturing sustainability risks in financial reporting, propose amendments in the CRR and CRD, as well as in the Solvency II Directive, to introduce climate stress tests for banks and scenario analyses for insurers, and will assess the suitability of macro-prudential tools to address climate-related risks and environmental risks more broadly.

Action 4 aims at increasing the contribution of the financial sector to sustainability by improving science-based target setting, disclosure and monitoring of the financial sector's commitments; clarifying the fiduciary duties and stewardship rules of investors to reflect the financial sector's contribution to the Green Deal targets; and improving the availability, integrity and transparency of ESG market research and ratings. Better financial institutions' disclosures of sustainability targets will be achieved in the context of the CSRD proposal and building on the regulatory technical standard under the SFDR (see section "Sustainability-Related Disclosures"). With respect to the fiduciary duties of investment and pension funds, EIOPA will assess the potential need to broaden the concept of the 'long-term best interest of members and beneficiaries' and introduce the obligation to consider sustainability impacts. Finally, the ESAP (see section "Sustainability-Related Disclosures") will make ESG data accessible, harmonized and machine-readable, contributing to improve the reliability, comparability and transparency of ESG ratings.

Action 5 addresses the financial system's transition and the issue of greenwashing. With respect to the latter, current supervisory powers, capabilities and obligations will be reviewed to make them adequate for addressing greenwashing. Concerning the transition of the financial system, a monitoring system will be developed, including the assessment of the additional financial investments needed to achieve the EU climate and environmental goals, and of the progress made in each individual Member State. Notably, Action 5 also includes the establishment of a Sustainable Finance Research Forum, to promote knowledge sharing between the academia and the finance industry on substantive sustainable finance issues.[56]

Finally, Action 6 deals with the international dimension of sustainable finance. In particular, it aims at promoting an ambitious consensus in international forums, e.g. on the concept of double materiality and disclosure frameworks (see section "Sustainability-Related Disclosures"), as well as on objectives and principles for taxonomies. It also includes the proposal to advance and deepen the work of the International Platform on Sustainable Finance (IPSF), which was established in 2019 to promote integrated markets for sustainable finance.[57] Finally, it includes the development of a strategy to support low- and middle-income countries in scaling up their access to sustainable finance, including by accelerating private financial flows towards EU partner countries.

Science-Based Regulation

Financial regulation for sustainable finance is rapidly evolving in the European policy landscape. However, it faces several challenges, and just like with a many-headed hydra, every time policymakers cut off the head of one problem, new problems grow instantly in its place. In this context, science can and should be a powerful ally. Indeed, while the crucial role of research in informing policy-making should be reaffirmed in general, this is all the more needed in the field of sustainable finance, where technical knowledge is essential, for example, for the definition of green and harmful activities and for the analysis of climate-related disasters. However, the contribution that policy-relevant research can and should give is not limited to environmental science.

For example, sections "Sustainability-Related Disclosures" and "The Delegated Legislation Integrating Sustainability into UCITS Directive and AIFMD" discussed the Taxonomy Regulation and the associated disclosure obligations, which will enter into force in a context where no sustainability disclosure obligations exist for SMEs, governments and central banks, as well as for non-EU firms. Against this background, policymakers and the finance industry need to find a solution to assess the Taxonomy-alignment of exposures towards these counterparts. To address this issue, several policy and academic studies have used the methodology developed by the Joint Research Centre of the European Commission in collaboration with the University of Zurich.[58] This approach, which is so far available for climate change mitigation, entails the use of standardized coefficients associated with the various

NACE (i.e. the statistical classification of economic activities in the European Community) sectors covered by the Taxonomy and can be used to estimate the Taxonomy-alignment of financial portfolios when no further information is available other than the NACE sector of the issuer/borrower.

With respect to green financial instruments, although green financial markets are growing exponentially, the incentive structure and decision-making processes of relevant economic actors are still largely unclear. However, these aspects are crucial to design policies and tools that can effectively redirect investment towards green activities. With reference to green bonds, for example, it has been shown that green bond issuers experience lower financing costs, but only if they are comparatively more 'credible', for example, thanks to third-party verification.[59] These results suggest that the overall impact of the additional costs entailed by the verification and assessment requirements included in the EU Green Bond Standard proposal could be lower than expected or even non-existing.

In a market where greenwashing is still widespread and tools to measure the actual environmental and social impact of financial investments still need to be developed, an important question relates to the effectiveness of the various sustainable financial products and instruments. In the case of green bonds, it has been shown that they are a credible signal of an issuer's environmental engagement, as green issuances are associated with a reduction in emissions.[60]

Another relevant aspect is linked to whether and how investors price sustainability risks. In this respect, it has been shown that European investors do price in climate risk (at least to a certain extent, that is becoming wider and wider; see Chapter 3). However, in order to assess the 'greenness' of a company they do not only look at its carbon emissions but also at the quality of its environmental disclosures.[61] This finding provides strong support for even more ambitious sustainability disclosure provisions. The determinants of relevant investment decisions are also still largely unknown. In this respect, scientific evidence documents a drop in the price of equities issued by highly polluting firms after the first global climate strike (Ramelli et al., 2020), and a reduction of 'brown' exposures in European investors' portfolios after the Paris Agreement (Alessi, Battiston and Kvedaras, 2021). Based on the findings of these and other papers, it is clear that climate risk is a factor with a significant market impact, which could possibly have financial stability consequences.

Indeed, with respect to climate-related financial risks, competent authorities and the industry all agree that more research is needed to develop models for the estimation of these risks and the design of actionable scenarios. One policy option on the table is the potential calibration of prudential requirements for banks, based on environmental considerations. On this point, in particular, scientific evidence is needed as soon as possible to inform regulators, policymakers and supervisors on how green and 'dirty' exposures should be treated, to better reflect their risk profile (Alessi et al., 2021). An essential building block for this line of research is of course the availability of reliable estimates of transition risks in financial portfolios (Alessi & Battiston, 2021). Models are also needed to quantify the exposure of financial institutions to physical risk, which is particularly complex owing to the need of matching geolocalized data on losses and risks with disaggregated financial data on exposures (Di Girolamo et al., 2021).

These are just few examples of open issues where policy would benefit from closer cooperation with the academia. At the same time, it would be helpful if researchers could focus their attention on the most pressing issues faced by the industry and the legislator, and increase their efforts towards clarifying the policy implications of their findings.

Notes

1. See Commission staff working document 'Identifying Europe's recovery needs' accompanying the document 'Communication: Europe's moment: Repair and Prepare for the Next Generation', European Commission, COM/2020/456 final, 27 May 2020.
2. The reported estimates are based on the previous target emission cut of 40%.
3. 'Financing a sustainable European economy', Report by the EU High-level expert group on sustainable finance, 2018.
4. 'Communication—Action Plan: Financing sustainable growth', European Commission COM(2018) 97 final, 8 March 2018.
5. 'Proposal for a Regulation of the European Parliament and of the Council on European green bonds', European Commission (COM(2021) 391 final, 6 July 2021).
6. 'Report on EU Green Bond Standard', Technical expert group on sustainable finance, 2019. 'Usability Guide—EU Green Bond Standard', Technical expert group on sustainable finance, 2020.

7. 'Development of EU Ecolabel criteria for Retail Financial Products', European Commission—Joint Research Centre, Technical Report, version 4.0, March 2021.
8. Undertakings for the Collective Investment in Transferable Securities (UCITS) and retail Alternative Investment Funds (AIFs).
9. Commission Delegated Directive amending Directive 2010/43/EU as regards the sustainability risks and sustainability factors to be taken into account for Undertakings for Collective Investment in Transferable Securities (UCITS), C(2021) 2617 final, 21 April 2021.

   Commission Delegated Regulation amending Delegated Regulation (EU) No. 231/2013 as regards the sustainability risks and sustainability factors to be taken into account by Alternative Investment Fund Managers, C(2021) 2615 final, 21 April 2021.

   Commission Delegated Regulation amending Delegated Regulations (EU) 2017/2358 and (EU) 2017/2359 as regards the integration of sustainability factors, risks and preferences into the product oversight and governance requirements for insurance undertakings and insurance distributors and into the rules on conduct of business and investment advice for insurance-based investment products, C(2021) 2614 final, 21 April 2021.

   Commission Delegated Directive amending Delegated Directive (EU) 2017/593 as regards the integration of sustainability factors into the product governance obligations, C(2021) 2612 final, 21 April 2021.

   Commission Delegated Regulation amending Delegated Regulation (EU) 2015/35 as regards the integration of sustainability risks in the governance of insurance and reinsurance undertakings, C(2021) 2628 final, 21 April 2021.

   Commission Delegated Regulation amending Delegated Regulation (EU) 2017/565 as regards the integration of sustainability factors, risks and preferences into certain organisational requirements and operating conditions for investment firms, C(2021) 2616 final, 21 April 2021.
10. Regulation (EU) 2019/2089 of the European Parliament and of the Council of 27 November 2019 amending Regulation (EU) 2016/1011 as regards EU Climate Transition Benchmarks, EU Paris-aligned Benchmarks and sustainability-related disclosures for benchmarks (*OJ* L 317, 9.12.2019, pp. 17–27).
11. Implementing and Delegated Acts: (i) Commission Delegated Regulation (EU) 2020/1816 of 17 July 2020 supplementing Regulation (EU) 2016/1011 of the European Parliament and of the Council as regards the explanation in the benchmark statement of how environmental, social and governance factors are reflected in each benchmark provided and published (*OJ L* 406, 3.12.2020, pp. 1–11); (ii) Commission Delegated Regulation (EU) 2020/1818 of 17 July 2020 supplementing Regulation (EU) 2016/1011 of the European Parliament and of the Council as

regards minimum standards for EU Climate Transition Benchmarks and EU Paris-aligned Benchmarks (*OJ L* 406, 3.12.2020, pp. 17–25); (iii) Commission Delegated Regulation (EU) 2020/1817 of 17 July 2020 supplementing Regulation (EU) 2016/1011 of the European Parliament and of the Council as regards the minimum content of the explanation on how environmental, social and governance factors are reflected in the benchmark methodology (*OJ L* 406, 3.12.2020, pp. 12–16).

12. 'Final Report—Guidelines on Disclosure Requirements Applicable to Credit Ratings', 18 July 2019, ESMA33-9-320.
13. 'Study on sustainability-related ratings, data and research', European Commission—Directorate-General for Financial Stability, Financial Services and Capital Markets Union and ERM, 2021.
14. Letter to Ms Mairead McGuinness, Commissioner in charge of Financial services, financial stability and Capital Markets Union, European Commission, 28 January 2021, ESMA30-379-423.
15. 'EBA action plan on sustainable finance', EBA, 6 December 2019.
16. 'EBA Discussion paper on management and supervision of ESG risks for credit institutions and investment firms', 30 October 2020, EBA/DP/2020/03.
17. 'Consultation Paper—Draft Implementing Standards on prudential disclosures on ESG risks in accordance with Article 449a CRR', 1 March 2021, EBA/CP/2021/06.
18. 'Opinion on Sustainability within Solvency II', 30 September 2019, EIOPA-BoS-19/241.
19. 'EBA Report on undue short-term pressure from the financial sector on corporations', EBA, 18 December 2019. 'Report—Undue short-term pressure on corporations', ESMA, 18 December 2019, ESMA30-22-762. 'Potential undue short-term pressure from financial markets on corporates: Investigation on European insurance and occupational pension sectors—Search for evidence', EIOPA, 18 December 2019, EIOPA-BOS-19-537.
20. Proposal for a Directive of the European Parliament and of the Council on Corporate Sustainability Due Diligence and amending Directive (EU) 2019/1937, COM/2022/71 final, 23 February 2022.
21. Regulation (EU) 2020/852 of the European Parliament and of the Council of 18 June 2020 on the establishment of a framework to facilitate sustainable investment, and amending Regulation (EU) 2019/2088 (*OJ L 198*, 22.6.2020, pp. 13–43).
22. 'EU Taxonomy Climate Delegated Act', 'Annex 1 to the EU Taxonomy Climate Delegated Act', 'Annex 2 to the EU Taxonomy Climate Delegated Act', C(2021) 2800/3.
23. See Canfora, P., Dri, M., Polidori, O., Solzbacher, C. and Arranz Padilla, M. (2021) 'Substantial contribution to climate change mitigation—a

framework to define technical screening criteria for the EU taxonomy', European Commission—Joint Research Centre, Technical Report.

24. 'Technical assessment of nuclear energy with respect to the "do no significant harm" criteria of Regulation (EU) 2020/852 ("Taxonomy Regulation")', European Commission—Joint Research Centre, Technical Report, Ref. Ares(2021)1988129—19/03/2021.
25. 'SCHEER review of the JRC report on technical assessment of nuclear energy with respect to the "do no significant harm" criteria of Regulation (EU) 2020/852 ("Taxonomy Regulation")', Scientific Committee on Health, Environmental and Emerging Risks, 2 July 2021. 'Opinion on the Joint Research Centre's Report Technical assessment of nuclear energy with respect to 'do no significant harm' criteria of Regulation (EU) 2020/852 ("Taxonomy Regulation")', Group of Experts on radiation protection and waste management under Article 31 of the Euratom Treaty, 2 July 2021.
26. Commission Delegated Regulation amending Delegated Regulation (EU) 2021/2139 as regards economic activities in certain energy sectors and Delegated Regulation (EU) 2021/2178 as regards specific public disclosures for those economic activities, C/2022/0631 final, 9 March 2022.
27. 'European Council meeting (10 and 11 December 2020)—Conclusions', European Council, EUCO 22/20, 11 December 2020.
28. Canfora P., Arranz Padilla M., Polidori O., Ostojic S., Dri M. (2022) 'Development of the EU Sustainable Finance Taxonomy—A framework for defining substantial contribution for environmental objectives 3–6', European Commission—Joint Research Centre, Technical Report.
29. In March 2022 the Sustainable Finance Platform published a report with recommendations on technical screening criteria for the remaining four environmental objectives of the EU taxonomy.
30. Bloomberg Industry Classification System and Global Industry Classification Standard (developed by MSCI and S&P), respectively.
31. In March 2022 the Sustainable Finance Platform published a report on Taxonomy extension options linked to environmental objectives to support the economic transition.
32. In February 2022 the Sustainable Finance Platform published a report on a Social Taxonomy, see https://ec.europa.eu/info/sites/default/files/business_economy_euro/banking_and_finance/documents/280222-sustainable-finance-platform-finance-report-social-taxonomy.pdf.
33. Directive 2014/95/EU of the European Parliament and of the Council of 22 October 2014 amending Directive 2013/34/EU as regards disclosure of non-financial and diversity information by certain large undertakings and groups (*OJ L* 330, 15.11.2014, pp. 1–9).

34. 'Communication—Guidelines on non-financial reporting (methodology for reporting non-financial information)', European Commission, C/2017/4234 (*OJ C* 215, 5.7.2017, pp. 1–20).
35. 'Communication—Guidelines on non-financial reporting: Supplement on reporting climate-related information', European Commission, C/2019/4490 (*OJ C* 209, 20.6.2019, pp. 1–30).
36. 'Proposal for a Directive of the European parliament and of the Council amending Directive 2013/34/EU, Directive 2004/109/EC, Directive 2006/43/EC and Regulation (EU) No. 537/2014, as regards corporate sustainability reporting', European Commission, 2021/0104 (COD), 21 April 2021.
37. A first set of draft EU sustainability reporting standards has been published for consultation by the EFRAG in April 2022.
38. Regulation (EU) 2019/2088 of the European Parliament and of the Council of 27 November 2019 on sustainability-related disclosures in the financial services sector, (*OJ L* 317, 9.12.2019, pp. 1–16).
39. Commission Delegated Directive amending Directive 2010/43/EU as regards the sustainability risks and sustainability factors to be taken into account for Undertakings for Collective Investment in Transferable Securities (UCITS); Commission Delegated Regulation amending Delegated Regulation (EU) No. 231/2013 as regards the sustainability risks and sustainability factors to be taken into account by Alternative Investment Fund Managers; Commission Delegated Regulation amending Delegated Regulations (EU) 2017/2358 and (EU) 2017/2359 as regards the integration of sustainability factors, risks and preferences into the product oversight and governance requirements for insurance undertakings and insurance distributors and into the rules on conduct of business and investment advice for insurance-based investment products; Commission Delegated Directive amending Delegated Directive (EU) 2017/593 as regards the integration of sustainability factors into the product governance obligations; Commission Delegated Regulation amending Delegated Regulation (EU) 2015/35 as regards the integration of sustainability risks in the governance of insurance and reinsurance undertakings; Commission Delegated Regulation amending Delegated Regulation (EU) 2017/565 as regards the integration of sustainability factors, risks and preferences into certain organisational requirements and operating conditions for investment firms.
40. 'Final Report on draft Regulatory Technical Standards with regard to the content, methodologies and presentation of disclosures pursuant to Article 2a(3), Article 4(6) and (7), Article 8(3), Article 9(5), Article 10(2) and Article 11(4) of Regulation (EU) 2019/2088', Joint Committee of the European supervisory Authorities, 2 February 2021, JC 2021 03.

41. See recital 5 Regulation (EU) 2019/2088 and EC explanatory memorandum (COM(2018) 354 final. 2018/0179(COD).
42. See ESMA final report on technical advice to the European Commission on integrating sustainability risks and factors into MiFID II (ESMA35-43-1737) and Delegated Act mentioned in Note 1.
43. 'Delegated act specifying the content and presentation of information to be disclosed by undertakings subject to Articles 19a or 29a of Directive 2013/34/EU concerning environmentally sustainable economic activities, and specifying the methodology to comply with that disclosure obligation', European Commission (C(2021) 4987 final, 6 July 2021).
44. Regulation (EU) 2019/876 of the European Parliament and of the Council of 20 May 2019 amending Regulation (EU) No. 575/2013 as regards the leverage ratio, the net stable funding ratio, requirements for own funds and eligible liabilities, counterparty credit risk, market risk, exposures to central counterparties, exposures to collective investment undertakings, large exposures, reporting and disclosure requirements, and Regulation (EU) No. 648/2012 (*OJ L* 150, 7.6.2019, pp. 1–225).
45. Regulation (EU) 2019/2033 of the European Parliament and of the Council of 27 November 2019 on the prudential requirements of investment firms and amending Regulations (EU) No. 1093/2010, (EU) No. 575/2013, (EU) No. 600/2014 and (EU) No. 806/2014 (*OJ L* 314, 5.12.2019, pp. 1–63).
46. Proposal for a Regulation of the European Parliament and of the Council establishing a European single access point providing centralised access to publicly available information of relevance to financial services, capital markets and sustainability, COM/2021/723 final, 25 November 2021.
47. The 'Financial sustainable package' is an ambitious and comprehensive package of measures to help improve the flow of money towards sustainable activities across the European Union.
48. Specifically European Commission, in June 2020, has published two consultations documents—on the basis of the technical advice to the Commission published by ESMA on 30 April 2019—proposing amendment to the Commission Directive 2010/43/EU (implementing Directive 2009/65/EC) and to the Delegated Regulation (EU) 231/20133 (supplementing Directive 2011/61/EU), in order to clarify requirements for management companies concerning the integration of sustainability risks and sustainability matters.
49. The definition of sustainability risk and the definition of sustainability factors laid down by the Regulation (EU) 2019/2088 on sustainability-related disclosures in the financial services sector are also recalled in the amendments to the UCITS and AIFMD Delegated Acts.
50. As far as the principal adverse impacts of investment decisions on sustainability factors are concerned, it is worth noting that the regulator has

envisaged a so called 'comply or explain approach', so that, on the basis of art. 4 of Disclosure Regulation, market participants—where they do not consider adverse impacts—shall publish and mantain on their websites clear reasons for why they do not do so, including relevant information as to whether and when they intend to consider such adverse impacts.

51. The ESAs, on 23 April 2020, has published a consultation document of Regulatory Technical Standards (RTS). The final version of RTS was published on 2 February 2021.
52. 'Communication: Strategy for Financing the Transition to a Sustainable Economy', European Commission, (COM(2021) 390 final, 6 July 2021).
53. 'Communication: EU Taxonomy, Corporate Sustainability Reporting, Sustainability Preferences and Fiduciary Duties: Directing finance towards the European Green Deal', European Commission, COM(2021) 188 final, 21 April 2021.
54. To improve the availability of data on natural disasters, the new EU Strategy on Adaptation to Climate Change designates the JRC Risk Data Hub (https://drmkc.jrc.ec.europa.eu/risk-data-hub#/) as the reference platform for the standardization of the recording and collection of comprehensive and granular climate-related losses and physical climate risk data at the EU level. To reach this objective, public–private partnership to collect and share such data will be encouraged, especially at subnational level. Based on the Risk data Hub, a common terminology and common metrics will be established.
55. See Chapter 9 for a discussion of other sustainable corporate governance initiatives, such as the 'European Parliament resolution of 10 March 2021 with recommendations to the Commission on corporate due diligence and corporate accountability'.
56. The JRC is leading on this initiative, as it involves strengthening sustainable finance research (see also Section "Science-Based Regulation") and knowledge transfer.
57. The IPSF currently includes 17 member jurisdictions, accounting for 55% of global GHG emissions and GDP, and 50% of world population.
58. Alessi, L., and Battiston, S. (2021). 'Two sides of the same coin: Green Taxonomy alignment versus transition risk in financial portfolios', Working papers in Economics and Finance 2021–14, European Commission—Joint Research Centre.
59. Fatica, S., Panzica, R., and Rancan, M. (2021). The pricing of green bonds: are financial institutions special? *Journal of Financial Stability*, Vol. 54.
60. Fatica, S., and Panzica, R. (2021). 'Green bonds as a tool against climate change?' *Business Strategy and the Environment*, Vol. 30, issue 5.

61. Alessi, L., Ossola, E., and Panzica, R. (2021). 'What greenium matters in the stock market? The role of greenhouse gas emissions and environmental disclosures', *Journal of Financial Stability*, Vol. 54.

References

Alessi, L., & Battiston, S. (2021). *Two sides of the same coin: Green taxonomy alignment and transition risk in financial portfolios*. Working papers in Economics and Finance 2021-14, European Commission—Joint Research Centre.

Alessi, L., Battiston, S., & Kvedaras, V. (2021). *Over with carbon? Investors' reaction to the Paris Agreement and the US withdrawal*. Working papers in Economics and Finance 2021-12, European Commission—Joint Research Centre.

Alessi, L., Battiston, S., Melo, A. S., & Roncoroni, A. (2019). *The EU sustainability taxonomy: A Financial Impact Assessment*. European Commission—Joint Research Centre, Technical Report.

Alessi, L., Di Girolamo, F., Pagano, A., & Petracco, A. (2021). *Accounting for climate transition risk in banks' capital requirements*. Working papers in Economics and Finance, European Commission—Joint Research Centre, forthcoming.

Alessi, L., Ossola, E., & Panzica, R. (2021). What greenium matters in the stock market? The role of greenhouse gas emissions and environmental disclosures. *Journal of Financial Stability, 54*, 100869.

Canfora, P., Arranz Padilla, M., Polidori, O., Ostojic, S., & Dri M. (2022). *Development of the EU sustainable finance taxonomy—A framework for defining substantial contribution for environmental objectives 3–6*. European Commission—Joint Research Centre, Technical Report.

Canfora, P., Dri, M., Polidori, O., Solzbacher, C., & Arranz Padilla, M. (2021). *Substantial contribution to climate change mitigation—A framework to define technical screening criteria for the EU taxonomy*. European Commission—Joint Research Centre, Technical Report.

Di Girolamo, F., Joossens, E., Petracco-Giudici, M., Pagano, A., & Schoutens, W. (2021). *Climate physical risk: impact on insurers' defaults and public finances*. Working Paper, European Commission—Joint Research Centre, forthcoming.

Fatica, S., & Panzica, R. (2021). Green bonds as a tool against climate change? *Business Strategy and the Environment*, *30*(5), 2688–2771.

Fatica, S., Panzica, R., & Rancan, M. (2021). The pricing of green bonds: are financial institutions special? *Journal of Financial Stability, 54*, 100873.

Ramelli, S., Ossola, E., & Rancan, M. (2020). Stock price effects of climate activism: Evidence from the first Global Climate Strike. *Journal of Corporate Finance*, *69*, 102018.

CHAPTER 9

Future Perspectives on Sustainable Corporate Governance and Disclosures

Alessandro D'Eri and Valerio Novembre

Corporate Disclosures

In Chapter 4 we have discussed the aspects relating to the application of the Non-Financial Reporting Directive (NFRD) and have emphasised multiple areas in which the existing requirements can be improved, most notably in the areas relating to: the scope of application, the location

The opinions stated by the authors are personal and do not bind ESMA.

A. D'Eri (✉)
European Securities and Markets Authority, Paris, France
e-mail: alessandro.deri@esma.europa.eu

V. Novembre
European Securities and Markets Authority, Paris, France
e-mail: valerio.novembre@esma.europa.eu

N. Linciano et al. (eds.), *Information as a Driver of Sustainable Finance*, Palgrave Studies in Impact Finance,
https://doi.org/10.1007/978-3-030-93768-3_9

of the non-financial statement, the notion of materiality, the standardisation and digitisation of the contents and the level of assurance and enforcement.

At the time of writing this chapter, the European Commission has issued a proposal for a new Corporate Sustainability Reporting Directive (CSRD) which will replace the NFRD which aims at addressing the concerns arising from the application of the current requirements and which has the potential to improve the quality and level of detail of the non-financial disclosures to also better support the information needs of financial market participants which are subject to the disclosure requirements of the Regulation on sustainability-related disclosures in the financial services sector (SFDR).

The CSRD proposal is not the only intervention that is likely to change the landscape of corporate disclosures in the next years. The July 2021 European Commission's Delegated Acts[1] specifying how financial and non-financial undertakings shall fulfil the disclosure requirements provided for by Article 8 of the Taxonomy Regulation have introduced additional disclosures within the non-financial statements. According to these requirements, in fact, the non-financial statements prepared in accordance with the NFRD shall include selected KPIs—i.e. turnover, capital expenditures and operational expenditures—providing information on the degree to which the activities carried on by the reporting entities are aligned with the EU Taxonomy for environmentally sustainable economic activities.

The CSRD proposals and the additional disclosures on the degree of taxonomy alignment could make the non-financial statements key documents to provide a comprehensive view on the sustainability profile of an entity. However, the future usefulness of the non-financial statements will largely depend on whether the key deficiencies identified in the current practice are addressed and whether, at the same time, the new requirements, including potentially any disclosures standards, will be flexible enough to adapt to a rapidly changing environment such as the one characterising sustainable finance.

In considering how to deliver more relevant, reliable and comparable non-financial disclosures, in February 2020 the European Commission published a questionnaire to collect input from stakeholders regarding the possible areas of the NFRD that require improvement. In summarising the feedback received from 588 organisations, the Commission noted that the vast majority of respondents and most notably users of non-financial information, consider that the status quo is unsatisfactory. The majority of preparers responding to the consultation also considered that additional requests for non-financial information, for example from rating agencies or NGOs, are a significant problem and more than a third complained about the current complexity and the difficulty in deciding what to disclose.

This feedback already provides a good indication of the widespread acceptance that the current requirements are not capable of supporting the increasing demand for quality sustainability-related disclosures. While there is a shared understanding on the shortcomings of the current reporting framework, different possible scenarios can be pursued to revise the NFRD and improve the level of relevance, reliability and comparability of corporate non-financial disclosures. In the rest of this chapter, we will address some of the possible developments of the non-financial reporting landscape and, in particular, we will focus on the CSRD proposals and consider how these reflect the input arising from the Commission's consultation on the revision of the NFRD and considering the most recent developments at European and international level in this area. At the time of finalising this chapter, the CSRD proposal is undergoing political negotiations which may result in changes compared to the Commission proposal which is largely the basis for the considerations set out in the following sections.

The Scope of Application of the Non-financial Reporting Requirements

As explained in Chapter 4, the NFRD currently requires disclosure from a fairly narrow population of large entities, i.e. large public-interest entities with more than 500 employees. The existing requirements have also given rise to diversity in practice given the divergent transposition of the Directive at national level, with some countries lowering the threshold to 250 employees.

In this context, it is not surprising that the vast majority of respondents to the Commission's consultation supported an expansion of the scope of the NFRD, but then different views as to how the scope should be changed were expressed. Most stakeholders indicated that all EU-listed companies, regardless of their size, should be in the scope of the Directive and that also those that are listed in the EU, but established in a third-country should be covered by the NFRD requirements. A significant proportion of respondents to the NFRD consultation also indicated support for including within the scope of the large non-listed entities.

As discussed in Chapter 4, the Accounting Directive, as well as other areas of EU legislation, are built around dimensional requirements of the concerned entities. While size of the reporting entities is generally regarded as an effective proxy of the resource constraint that affects companies, it is not equally effective as a proxy of their sustainable footprint.

While the expansion of the scope of the NFRD seems to be a necessary step in the direction of ensuring that the entities that are subject to the ESG disclosure obligations in other areas of sustainable finance legislation (such as the SFDR, the Taxonomy Regulation and the Benchmark Regulation) will have the necessary data points to prepare their disclosures, the question of how to address proportionality concerns of smaller entities remains a relevant one. One possible alternative would be to consider a combination of size and sector to identify a proxy not only of resource constraints, but also of sustainability impact.

More in general, it would seem reasonable to expect that some form of simplified reporting standard be made available following the revision of the NFRD for Small and Medium-sized entities (SMEs). The question is then whether such simplified reporting regime should be mandatory or voluntary. In this respect, some respondents to the NFRD consultation suggested distinguishing between listed SMEs in high-risk sector from a sustainability perspective for which the reporting requirements should always apply. For the other SMEs, including some listed ones which do not operate in high-risk sectors, the reporting regime could be simplified. In this respect, it should also be noted that a trade-off exists between the two principles of proportionality and investor protection when different transparency requirements may be applied for listed issuers of different size (e.g. more detailed requirements for larger entities than for smaller ones).

As the European efforts towards mainstreaming sustainability in financial markets progress, SMEs will inevitably be required to provide evidence of their sustainability performance and impact, just like, for example, they provide basic financial metrics in order to get access to bank-based financings, but also—and perhaps most importantly—in the context of the European initiative on the Capital Markets Union (CMU) which aims at expanding the access to market-based finance to a wider audience of potential issuers.

The CRSD proposals address these issues by foreseeing an enlargement of the scope of application of the new sustainability reporting requirements—the CSRD no longer uses the term "non-financial statement"—to include all entities that are "large" under the meaning of the Accounting Directive and all entities that are listed on regulated markets, including small and medium ones (and except for micro-entities), as well as all banks and insurance undertakings including those that are incorporated in the form of mutuals and cooperatives provided when they meet certain dimensional requirements. The expansion of the scope to listed SMEs is envisaged with a time lag of three years, compared to the entry into force of the new proposals for the other entities, that is 2023. Non-listed SMEs are, however, given the possibility of applying the new regime on a voluntary basis. The CSRD proposals also clarify that third-country issuers listed on an EU regulated market shall comply with the new sustainability reporting requirements and an equivalence regime may be established to that effect in due course. As for all proposed requirements, the final outcome of the political negotiations may favour a different set of scoping requirements most notably in the area of SMEs. Similarly, it can be expected that, given the need to allow for sufficient time to implement the new requirements, the date of initial application of the new requirements may shift forward by one year, i.e. in 2024.

The Location of the Non-financial Information

Another important aspect that is often cited in the context of the future development of non-financial reporting is whether or not the non-financial statement shall be mainly—or solely—intended to form an integral part of the management report or be hosted in a separate report. In addition, the question on the location of the information is also relevant for the corporate governance statement and the remuneration report for which a similar regime as the one for the non-financial statement applies with respect to its inclusion either as part of the management report or included in a separate report (see section "Non–financial Reporting in the Context of Periodic Financial Information" in Chapter 4).

In the digital age, this question might sound rather basic given that data consumption is increasingly performed by means of IT tools and therefore the *physical* location of the disclosures is, from this perspective, of limited relevance. Algorithms appropriately set may identify the necessary information irrespective of the specific silo in which it is hosted.

However, *machine-readability* of the disclosures does not—and should not—be a reason for depriving users of another important piece of information, i.e. context. In fact, extracting specific data points from an entire set of disclosures in order to make them available in a standardised format to help financial analysis, does not imply that such analysis will not also benefit from putting such data points within the broader context of the company report which they belong to. On top of this, one of the arguments often put forward by reporting entities to argue for more flexibility in reporting regimes is to allow them telling *their own story*. If users of corporate reports are to grasp this story, reading corporate reports—or parts thereof—as a booklet rather than as a mere collection of data points may provide more insights. Obviously, this does not mean that IT tools cannot help with putting selected data points into a broader context—this is obviously one of the key benefits of data mining techniques which rely on Artificial Intelligence—but this implies that the choice of the location of the non-financial information has not become irrelevant.

As explained in Chapter 4, the current NFRD allows for flexibility in this area, whereas the default option is to include the non-financial statement in the management report and the alternative option is to present it as a separate report. Such flexibility has raised concerns in practice given that it has resulted in diversity in the level of detail and quality of the disclosures provided and, in some cases, in the limitation or even the impossibility for national competent authorities to exercise supervision on the non-financial information published by entities.

Some view the non-financial information as constituting a separate information silo from the management report and the financial statements and, as such, requiring a separate document. A separate sustainability report would, in fact, allow for more detailed disclosures which would mirror on the non-financial side, the length and complexity of the financial information that is typically provided in management reports and financial statements. However, others see non-financial information as a necessary complement of financial information and therefore their location should reflect the need for stronger connectivity between these two types of information. In addition, the use of a separate report may be regarded as signalling that non-financial information is of less importance than financial information.

An intermediate approach could be regarded as the one proposed by Accountancy Europe in its “Core and More” approach whereas one core report would contain only the key information in relation to both financial and non-financial information and then the more detailed separate reports would provide additional information and context for different areas.

In this area, the CSRD proposals require the inclusion of the sustainability reporting within an entity’s management report. This approach may result in better connectivity between financial and sustainability disclosures across the annual financial report and it will therefore no longer allow for the possibility to prepare a separate sustainability report, unless such a standalone report is included within the management report. It shall be noted that this proposal on the location of the sustainability reporting will also have a beneficial effect on clarifying the perimeter of the supervision and enforcement obligations of national supervisors.

Materiality and Standardisation

The double materiality perspective provided by the NFRD is a cornerstone of the sustainability reporting regime. The vast majority of respondents agreed with this concept and also acknowledged that alone, the materiality notion contained in article 2 of the Accounting Directive would not suffice to provide the social and environmental materiality, but it would only be suitable to capture ESG information that is financially material.

What is clear from the current reporting practice is that not many entities explain their *thought process* in terms of application of the materiality principle and how this results in disclosures of non-financial information. Information in this area was regarded as important input by the respondents to the Commission’s consultation as well as information on *how to* apply the double materiality concept to the different types of non-financial matters.

In addition to the above-mentioned improvements to the disclosure regime, however, in developing the future framework for non-financial disclosures, it would be important to understand how the double materiality notion will interplay with the application of the materiality concept in the broader context of international standardisation.

A very clear signal in this respect has come in September 2020 with the initiative of five international organisations (CDP, CDSB, GRI, IIRC and

SASB) that are active in the area of non-financial disclosures and standardisation and which have signed a Statement of Intent[2] in which, among other things, they committed to join forces to outline an approach to achieve "globally agreed set of sustainability topics and related disclosure requirements that can serve distinct materiality concepts". This initiative was also welcomed by the Sustainability Task Force of the International Organisations for Securities Commissions (IOSCO).[3] The same organisations, in December 2020, proposed a prototype[4] for climate-related financial disclosures which focuses on the specific lens of enterprise value creation.

In the Statement of Intent, the five organisations outline an approach of building blocks based on distinctive materiality concepts. As further explained in the December 2020 publication, these organisations explain that different materiality notions would result in two main types of reporting or which can also be thought of as separate, but interoperable, building blocks: enterprise value reporting, which is comprised of sustainability-related financial disclosures and traditional financial accounting, and sustainability reporting more strictly speaking.

These forms of reporting reflect the use of different lenses through which it is possible to look at non-financial matters. Financial statements would depict directly monetary amounts reflecting the consequences at entity-level of ESG factors, such as physical and transition climate-related risks. Disclosures provided outside of the financial statements would typically address the value creation drivers of an entity and, from that perspective, consider how some sustainability matters may create or erode enterprise value. Finally, sustainability reporting would address "all sustainability matters that reflect significant positive or negative impacts on people, the environment and the economy".

These three lenses are described as *nested* and *dynamic*. They are nested because "the nature of disclosure on the same sustainability matter" (such as Scope 1 emissions) that serves each perspective can vary (e.g., from calculation of emissions relative to ecological thresholds to adjustment of net asset values) and because "the universe of sustainability matters covered by each form of reporting is not necessarily the same". These lenses are dynamic because "sustainability topics that a company once considered immaterial for disclosure can become material, based on evidence of an organisation's impacts on the economy, environment and/or people. Likewise, some of these sustainability topics can also become material for enterprise value creation, either gradually or rapidly".

These five international organisations propose that their prototype of climate-related financial disclosures focuses on enterprise value creation and that this prototype can be used as a basis for developing an international standard of broad acceptance. Such a standard could be set up ultimately by the IFRS Foundation which in September 2020 launched a public consultation on a possible role that the Foundation might have in the area of sustainability reporting. Following the broad support gathered by stakeholders to set up an International Sustainability Standards Board (ISSB), the IFRS Foundation has taken concrete steps to accommodate in its Constitution and governance structure the ISSB and consolidate and build on the work of international standards and framework providers, such as the Sustainability Accounting Standards Board (SASB), the International Integrated Reporting Framework and the Climate Disclosures Standards Board (CDSB), to develop a first set of sustainability-related financial disclosures with a climate-first focus. This is a first step in the process of developing a full set of sustainability-related reporting standards. The objective of the IFRS Foundation with the establishment of the new board—as announced on 3rd November 2021 by the Trustees of the Foundation—is not to cover the full spectrum of the sustainability-related information, but to focus on disclosures that meet the needs of investors globally when assessing enterprise value creation. This "global baseline" of sustainability-related standards will then be interoperable with the jurisdiction-level or regional initiatives that are focused on public policy objectives that are specific to the concerned jurisdictions and reasons. In other words, the ISSB will look at materiality from the perspective of enterprise value creation thus focusing only on the so-called financial materiality. On 31st March 2022, the ISSB launched a 120-day consultation on its first two Exposure Drafts setting out proposals respectively for General Requirements for Disclosure of Sustainability-related Financial Information and Climate-related Disclosures. Another important development in the area of financial materiality, is the launch by the U.S. Securities and Exchange Commission on 21st March 2022 of a consultation on proposals to Enhance and Standardize Climate-Related Disclosures for Investors, both inside and outside the financial statements of listed entities.

These international developments suggest that the compatibility and interoperability of disclosure requirements that are based on different notions of materiality will become the key issue to understand how the international disclosure landscape will look like in the coming years especially now that the CSRD proposal has reaffirmed the double materiality concept as the basis for the development of reporting standards which will be mandatorily applicable to the entities in scope (with a differentiation for SMEs for which a dedicated set of standards shall be developed).

This approach is rather consistent with the wide consensus in the Commission's public consultation on the double materiality principle and towards the development of mandatory reporting standards for non-financial disclosures together with, possibly, voluntary, simplified standards for SMEs.

The CSRD proposal has identified in the European Financial Reporting Advisory Group (EFRAG) the future standard setter to develop the draft reporting requirements which will then be submitted to the European Commission for their incorporation in specific delegated acts, after consultation with EU public authorities, most notably ESMA. On 29th April 2022 EFRAG launched a public consultation on its first set of draft European Sustainability Reporting Standards (ESRS). These standards address the full spectrum of ESG-related information as envisaged in the CSRD proposal.

Given the direction taken in the EU with the European Commission's proposal in the CSRD to require the adoption of EU-specific reporting standards, including proportionate ones that are specific for SMEs, there are still some open questions to address, for example: will the reporting standards be compatible with international initiatives so to enable future convergence? Will international standards be granted (at least partial) equivalence to the future ESRS in Europe? With the parallel development of EU-specific and international reporting standards, is there still room for a target scenario in which for Europe the international standards constitute an accepted baseline to which it is added a *top-up* of requirements addressing specific European policy objectives? Will the governance of the standard-setting process envisaged in the CSRD with EFRAG as the standard-setter prove to be the most appropriate to ensure high-quality standards?

The pace at which standard setting is moving and consolidating internationally with the establishment of two main standardsetters, i.e. the ISSB for 'enterprise value creation' and the Global Reporting Initiative for 'impact materiality'—seems to suggest that the global baseline ingredients exist to establish international standards that can effectively cover the EU double-materiality perspective. Such a scenario may only require Europe to add an EU-specific layer to reflect the specificities of requirements of European legislation in the area of sustainable finance. However, for a few years ahead, if finalised in its current form, the CSRD will foresee an EU-specific standard-setting regime that will be as compatible as possible with existing international initiative. The CSRD proposal, however, also envisages the possibility of reviewing the European standards at least every three years so to take into account any relevant developments including those in the field of international standardisation.

Digitisation of the Sustainability Reporting

Under the applicable requirements, there is no obligation to prepare the non-financial statements in a structured electronic format similar to the one required for the preparation of annual financial report on the basis of the so-called European Single Electronic Format (ESEF).

The large majority of the respondents to the NFRD consultation supported the tagging of non-financial information under the condition that this is done against reporting standards. Furthermore, most respondents emphasised the importance of making the non-financial information available through a single access point.

The CSRD proposal builds on this input to require the mandatory tagging of the sustainability reporting as well as of the disclosures required by Article 8 of the Taxonomy Regulation and expands the requirements to prepare the financial statements and management reporting in XHTML format pursuant to Article 3 of the Commission Delegated Regulation 2019/815 to all entities that are subject to the obligation to prepare a (consolidated or individual) sustainability report. This requirement implies that the future European standards will need to cater of a sustainability reporting digital taxonomy similar to what happens today for financial statements prepared under IFRS for which the IFRS Foundation provides a taxonomy to enable digital reporting.

Enforcement and Assurance

Finally, the control ecosystem surrounding the non-financial reporting requirements is currently far from the one applicable to other parts of an entity's annual financial report, most notably the management report and the financial statements. This is mainly due to the flexibility provided by the NFRD on the level of assurance necessary for the non-financial statements as well as the poor coordination between the NFRD and the Transparency Directive which results in different levels of oversight on the non-financial information envisaged by the varying national transposition measures of the NFRD. There is also a lack of clarity on the required involvement of the administrative body of the reporting entities on the non-financial statement.

The majority of respondents to the NFRD consultation indicated that differences across EU jurisdictions particularly in the assurance requirements between financial and non-financial information are not justifiable and the majority of respondents indicated also that stronger assurance requirements are needed for non-financial information, however, split views were expressed as to whether limited or reasonable assurance should be required on this type of information.

The CSRD proposal introduces a number of measures that have the potential to improve the quality of the control environment in relation to the sustainability reporting. Firstly, it proposes a mandatory limited assurance requirement which can be fulfilled by means of an opinion by either statutory auditors or audit firms as well as by independent assurance providers. Secondly, it proposes to require that in the responsibility statement that is included in the annual financial report of listed entities makes explicit reference to the fact that the people responsible for financial reporting in the issuer declare that the management report is prepared in accordance with the sustainability reporting standards and it further clarifies the responsibility of the administrative, management and supervisory bodies of listed and non-listed entities to ensure that the companies report on the basis of the sustainability reporting standards. Thirdly, it improves the coordination between the Accounting Directive and the Transparency Directive to explicitly include the sustainability reporting within the remit of the supervisors that are currently already responsible of the enforcement of the applicable requirements for the financial statements and the management reports of issuers.

With these measures, the CSRD proposal has the potential to promote more consistent and robust implementation of the sustainability reporting requirements and to improve the quality of reported information in this area.

Corporate Governance

From Corporate Disclosures to Governance: The Sustainable Corporate Governance Initiative

While policy work has initially focused on sustainable corporate disclosures, increasing interest has later emerged in the area of sustainable corporate governance. The European Green Deal announced in December 2019 already set out that "*sustainability should be further embedded into the corporate governance framework, as many companies still focus too much on short-term financial performance compared to their long-term development and sustainability aspects*".[5]

Following up to the European Green Deal, the Commission announced in the course of 2020 a Sustainable Corporate Governance Initiative for 2021. This initiative leveraged on the analytical work performed under Action 10 of the Commission's Action Plan on Financing Sustainable Growth and was intended to also be part of the Renewed Sustainable Finance Strategy.[6] Echoing the 2019 Green Deal, the Sustainable Corporate Governance Initiative was presented as a tool to ensure that sustainability would be further embedded into the corporate governance framework with a view to better align the long-term interests of management, shareholders, stakeholders and society. In the Commission's view, corporate boards should be able to fully integrate stakeholder interests, sustainability risks, opportunities and environmental, social, human and economic adverse impacts into their strategies, decisions and oversight role.

The Sustainable Corporate Governance Initiative is complementary to the review of the Non-Financial Reporting Directive (NFRD[7]), that led to the publication of the proposal for a Corporate Sustainability Reporting Directive (CSRD)[8] in April 2021. As illustrated in Chapter 4, consistently with the approach taken in the NFRD, the CSRD proposal requires companies to publicly disclose information both on how they are affected by non-financial issues and on their own impact on the society and the environment. Companies should disclose their social and environmental policies and their due diligence processes or they should explain why they do not have any via a comply or explain approach.

Although the CSRD is expected to clarify the requirement to report on due diligence processes, such requirement would still not be underpinned by a corporate obligation to carry out due diligence (and mitigate any adverse impact thereof). In this context, the Sustainable Corporate

Governance Initiative aims to complement the CSRD by introducing duties to act rather than only to report. By ensuring that reporting obligations are paired with adequate corporate and director duties, this initiative can be expected to enhance the reliability of the ESG information that is disclosed to the market, thereby holding companies fully accountable for how they mitigate their adverse environmental and social impacts.[9] Such concrete duties are in turn supposed to contribute to reducing the risk of "green-washing", thereby also supporting the objectives of the NFRD review.

The Analytical Work Behind the Sustainable Corporate Governance Initiative

Back in 2018, the High-Level Expert Group on Sustainable Finance[10] already recommended to clarify board members' duties with two goals in mind: (i) to provide them with better incentives to take into account stakeholder interests and (ii) to require them to adopt a sustainability strategy so to facilitate the integration of sustainability considerations in the corporate strategy. In the following 2018 Action Plan on Financing Sustainable Growth,[11] the Commission indicated that it would perform analytical work on the possible need to legislate in this area. As part of this work, two studies were carried out, one on directors' duties and one on due diligence requirements, both finding certain market failures and calling for EU action. These were later used by the Commission as a basis for its inception impact assessment on potential regulatory initiatives.

The Study on Directors' Duties

Prepared by the consultancy firm Ernst Young, the study on directors' duties and sustainable corporate governance finds that "*over the past three decades there has been a trend for EU listed companies to focus on the short-term benefit of shareholders rather than on the long-term interests of the company*". According to the report, data indicate an upward trend in shareholder pay-outs, which increased from 20 to 60% of net income while the ratio of CapEx and R&D spending to net income reportedly declined by 45 and 38% respectively.

On this basis the study argues that a number of factors, both enshrined in the regulatory framework and in market practice, have facilitated and gradually exacerbated corporate short-termism and that pressure to focus on short-term financial performance reduces companies' ability to integrate sustainability considerations adequately into business strategies and decisions. This, in turn, is seen as detrimental for the long-term development of the economy as it may curb the investments necessary to perform the sustainability transition and it is potentially conducive to higher income inequality.

The EY study identifies the following market and regulatory failures as key drivers of these problems. Firstly, the company interest and director duties that are defined in national law tend to be interpreted narrowly favouring short-term maximisation of shareholder value. Secondly, short-term investors' pressure on companies plays a role together with directors' remuneration structures that are linked to short-term financial performance. Thirdly, corporate governance frameworks do not sufficiently foster the interest of other stakeholders, including those affected by the company as well as the local and global environment.

Against these findings, the EY study puts forward a number of possible policy objectives that are seen as able to lengthen the time horizon in corporate decision-making and promote corporate governance that is more conducive to sustainability. In particular, it is suggested that the EU should: (i) strengthen the role of directors in pursuing their company's long-term interest; (ii) improve directors' accountability towards integrating sustainability into corporate strategy and decision-making; and (iii) promote additional corporate governance arrangements that contribute to company sustainability (e.g. in the area of board remuneration, board composition, etc.).[12]

The findings of the EY study have been criticised by a number of observers, especially academics, arguing in the first place that the evidence of increasing short-termism that is provided in the study comes from measures of gross corporate pay-outs that the literature has already criticised as not actually measuring short-termism.[13] In addition, these scholars[14] claim that while the study is framed in terms of short-termism, most of the unfavourable consequences that are identified are externalities and inequitable distribution problems. Thirdly, academics have stressed that shareholder value creation is not per se a short-term phenomenon, on the contrary the shareholder value model reflects future risk and opportunities to generating returns and as such it is able to evaluate the impact of environmental, social and governance factors on companies' risk-adjusted performance.[15] In their view, "*short-term investors can be an additional important corporate governance mechanism in combination with long-term investors and their incentive to exert influence on firm management through engagement*".[16]

The Study on Due Diligence Obligations

Again prepared as a follow up to Action 10 of the European Commission Action Plan on Financing Sustainable Growth, the study on due diligence obligations carries out analytical and consultative work with relevant stakeholders to "*assess the possible need to require corporate boards to develop and disclose a sustainability strategy, including appropriate due diligence throughout the supply chain, and measurable sustainability targets*".[17] The study starts from the assumption that companies may not be able to properly identify and address climate change and other environmental, social and human rights (including workers' rights, child labour etc.) risks and impacts in their operations and supply chains.

As a background, it should be noted that the concept of due diligence relevant for this study is broadly defined as the duty to "*identify, prevent, mitigate and account for*" potential and actual adverse corporate impact on human rights and the environment, either in the firms' own operations or through their supply and value chain. This concept was introduced by the UN Guiding Principles on Business and Human Rights[18] and incorporated into the OECD Guidelines for Multinational Enterprises[19] to extend to other areas of responsible business conduct such as the environment and climate change, conflict, labour rights, bribery and corruption, disclosure and consumer interests, as well as in the ILO Tripartite declaration of principles concerning multinational enterprises and social policy.[20]

The study, prepared by the British Institute of International and Comparative Law, Civic Consulting and LSE Consulting, reviews market practices across the EU to find evidence of market and regulatory failures affecting the current mostly voluntary framework. It then performs a regulatory analysis to identify and assess the possible policy options. It sets out the following ones: (i) No EU policy change, whereby current national laws are expected to continue to apply, requiring mandatory due diligence only in certain Member States with the possibility of environmental and human rights risks and impacts in companies' value chains being not sufficiently integrated into corporate strategies, (ii) New EU-level guidelines or guidance aimed at providing companies with further incentives to undertake due diligence through the supply chain, (iii) New EU regulation requiring due diligence reporting in further detail than that required by the NFRD/CSRD (and going beyond the NFRD/CSRD concept of materiality), and (iv) New regulation requiring companies to perform due diligence as a legal duty of care, with a number of sub-options, addressing the scope of companies potentially subject to the duty and the oversight/enforcement mechanisms.

EU Policy Actions Under the Sustainable Corporate Governance Initiative

Based on the above analytical work, the EU Commission was expected to assess which policy actions, if any, should be put forward as part of its Sustainable Corporate Governance Initiative. The exact type of actions would be identified based on the inception impact assessment and the related consultation process.

Published at the end of 2020, a public consultation document collected a broad range of views from stakeholders with regard to the Sustainable Corporate Governance Initiative.[21] It built on the data collected and conclusions reached by the two above-mentioned studies as well as on the feedback received to the public consultation on the Renewed Sustainable Finance Strategy.

In addition, the Commission published an inception impact assessment that sets out the following potential policy mechanisms:

- Companies' obligation to address their adverse sustainability impact such as climate change, environmental, human rights (including workers and child labour) harm in their own operations and in their value chain by identifying and preventing relevant risks and mitigating negative impacts (due diligence duty);
- Directors' duty to take into account stakeholders' interests which are relevant for the long-term sustainability of the firm or which are among those affected by it (employees, environment, other stakeholders affected by the business, etc.), as part of their duty of care to promote the interests of the company and pursue its objectives, and integrate those interests into the corporate strategy;
- Appropriate facilitating, enforcement and implementation mechanisms accompanying these duties, including possible remediation where necessary;
- Additional possible corporate governance arrangements (e.g. directors' remuneration etc.).

In light of the wide scope envisaged for the Sustainable Corporate Governance Initiative, the Commission clarified its intention to use both non-legislative and legislative measures. An EU-level initiative was seen as necessary to tackle issues that Member States' action alone is unlikely to address as sustainability and short-termism problems are of a global dimension and have cross-border effects. This would also respond to the European Parliament's (EP) push in this direction.[22] A Commission's legislative proposal was expected in the course of 2021 and to include a modification of the Company Law Directive[23] and/or of the SRDII,[24] as explicitly acknowledged in the inception impact assessment.

Differently, following repeated postponements, a proposal for an entirely new Directive on Corporate Sustainability Due Diligence[25] (CSDD) was published by the Commission at the time of sending this chapter to the press in early 2022. According to the introductory text to the proposal, "*the Directive is more focused and targeted compared to the preferred option outlined in the draft impact assessment. The core of it is the due diligence obligation, while significantly reducing directors' duties by linking them closely to the due diligence obligation*".[26]

The CSDD proposal aims to ensure that companies active in the EU market contribute to sustainability through the "*identification, prevention and mitigation, bringing to an end and minimisation of potential or actual adverse impacts*"[27] on human rights and the environment (including those connected to subsidiaries and established direct and indirect business relationships throughout their value chains). To this goal, companies should take appropriate steps, including by setting policies and procedures, that can be reasonably expected to prevent or minimise adverse impacts.[28]

The proposed Directive does not regulate directors' duties as such but clarifies that the general duty of care of directors should be applied in a manner which is consistent with the due diligence obligations it imposes, and more generally that when directors act in the interest of the company, they have a duty of care to take into account human rights, climate and environmental consequences of their decisions (and more broadly all "*consequences of their decisions for sustainability matters*").[29]

Specific due diligence duties are assigned to directors, who are responsible for putting in place and overseeing due diligence actions and in particular a due diligence policy and integrating those into the corporate strategy. In addition, in order to appropriately skew directors' incentives, it is provided that, when applicable, variable remuneration packages should take into account the company's sustainable business strategy—including the Paris agreement-compliant transition plan provided for by Art. 15—and directors' contribution to that.

Going forward, one element of utmost importance will be to carefully align this initiative with the final text of the CSRD as well as with the other parts of the regulatory framework in the area of sustainability such as the SFDR and the Taxonomy Regulation. While full alignment might not always be possible, efforts should be directed at avoiding at least major inconsistencies.

In this regard, it should also be noted that the proposed scope of the CSDD Directive is different from that envisaged in the CSRD proposal (see *supra* par. 2.1.1), notably in relation to SMEs. According to the CSDD proposal, rules would cover EU limited liability companies with more than 500 employees and a net turnover worldwide that is higher than 150m euros (as well as non-EU companies with turnover generated in the EU of at least 150m euros). At a later stage, it is envisaged that other EU limited liability companies operating in certain high-impact sectors (e.g. the textile, mining and agriculture industries) would fall in the scope, provided that they have more than 250 employees and a net turnover worldwide that is higher than 40m euros (or non-EU companies with turnover generated in the EU of at least 40m euros).

Finally, it should be noted that enforcement and implementation mechanisms to ensure compliance with CSDD corporate duties are explicitly provided for by the proposal, including in terms of supervisory authorities and their powers. Public enforcement mechanisms will need to be carefully designed by Member States and they may leverage on existing and consolidated supervisory frameworks, for example in the area of corporate disclosures, to facilitate a swift implementation of the new rules into consistent and robust market practices.

Notes

1. The European Commission's delegated acts relating to the disclosure obligations pursuant to Article 8 of the Taxonomy Regulation are available at the following link: https://ec.europa.eu/finance/docs/level-2-measures/taxonomy-regulation-delegated-act-2021-4987_en.pdf. At the time of writing this chapter these delegated acts are undergoing the scrutiny of the European Parliament and the Council.

2. https://29kjwb3armds2g3gi4lq2sx1-wpengine.netdna-ssl.com/wp-content/uploads/Statement-of-Intent-to-Work-Together-Towards-Comprehensive-Corporate-Reporting.pdf.
3. https://29kjwb3armds2g3gi4lq2sx1-wpengine.netdna-ssl.com/wp-content/uploads/Reporting-on-enterprise-value_climate-prototype_Dec20.pdf.
4. https://29kjwb3armds2g3gi4lq2sx1-wpengine.netdna-ssl.com/wp-content/uploads/Reporting-on-enterprise-value_climate-prototype_Dec20.pdf.
5. EC (2019), The European Green Deal—communication (https://eur-lex.europa.eu/resource.html?uri=cellar:b828d165-1c22-11ea-8c1f-01aa75ed71a1.0002.02/DOC_1&format=PDF), pg. 17.
6. EC (2018), Renewed sustainable finance strategy and implementation of the action plan on financing sustainable growth|European Commission (https://ec.europa.eu/info/publications/sustainable-finance-renewed-strategy_en).
7. Directive 2014/95/EU.
8. EC (2021), Proposal for a DIRECTIVE OF THE EUROPEAN PARLIAMENT AND OF THE COUNCIL amending Directive 2013/34/EU, Directive 2004/109/EC, Directive 2006/43/EC and Regulation (EU) No. 537/2014, as regards corporate sustainability reporting COM/2021/189 final. https://eur-lex.europa.eu/legal-content/EN/TXT/?uri=CELEX:52021PC0189
9. Here, it should be noted that while providing market participants with relevant ESG disclosures is found to increase non-financial performance (Clarkson et al., 2008; Lopez de Silanes et al., 2019), the academic literature is inconclusive on the correlation between non-financial performance and board characteristics. However, a recent meta study by Endrika et al. (2020) find that female board representation, board size and board independence are interrelated and jointly influence CSR. The integration of sustainability factors in companies' boards is found to facilitate shareholder engagement on ESG issues which, in turn, can positively influence companies' ESG rating. Cf. Endrikat, J., C. de Villiers, T. W. Guenther, and E. M. Guenther (2020), Board characteristics and corporate social responsibility: A meta-analytic investigation. *Business & Society*. https://doi.org/10.1177/0007650320930638.
10. HLEG (2018), Final report of the High-Level Expert Group on sustainable finance | European Commission (europa.eu).
11. EC (2018), Renewed sustainable finance strategy and implementation of the action plan on financing sustainable growth | European Commission (europa.eu).
12. EY (2019), pg. vii.

13. Law and Business Professors' Submission to the EU on EY's "Study on directors' duties and sustainable corporate governance", 8 October 2020, available at: Feedback from: Holger Spamann (europa.eu).
14. Ibid.
15. Feedback Statement from Hamburg University Professors on the Sustainable Corporate Governance Initiative of the Directorate-General for Justice and Consumers, October 2020, available at: Feedback from: University of Hamburg (europa.eu).
16. Ibid.
17. EC (2018), Renewed sustainable finance strategy and implementation of the action plan on financing sustainable growth | European Commission (europa.eu).
18. UN (2011), guidingprinciplesbusinesshr_en.pdf (ohchr.org).
19. OECD (2011), Guidelines—Organisation for Economic Co-operation and Development (oecd.org).
20. ILO (2017), Tripartite Declaration of Principles concerning Multinational Enterprises and Social Policy (MNE Declaration)—5th Edition (2017) (ENTERPRISES) (ilo.org).
21. The questionnaire targeted a broad range of stakeholders and contained 26 questions including the possibility of providing additional information through open-ended questions. In total, 473,461 public responses were obtained during the consultation period. One of the responses had 122,785 supporting citizen signatures. In addition, currently 149 position papers were received outside of the EU Survey. This number was driven to a large extent by campaigns (online consultation tools) carried out by a number of NGOs, using pre-filled questionnaires.
22. EP (2020), Resolution on sustainable corporate governance. According to such resolution, "the EP Calls on the Commission to present a legislative proposal to ensure that directors' duties cannot be misconstrued as amounting solely to the short-term maximisation of shareholder value, but must instead include the long-term interest of the company and wider societal interests, as well as that of employees and other relevant stakeholders; believes, in addition, that such a proposal should ensure that members of the administrative, management and supervisory bodies, acting within the competences assigned to them by national law, have the legal duty to define, disclose and monitor a corporate sustainability strategy". https://www.europarl.europa.eu/doceo/document/TA-9-2020-0372_EN.html
23. Directive (EU) 2017/1132 of the European Parliament and of the Council of 14 June 2017 relating to certain aspects of company law (Text with EEA relevance). (https://eur-lex.europa.eu/legal-content/EN/TXT/?uri=CELEX%3A32017L1132).

24. Directive (EU) 2017/828 of the European Parliament and of the Council of 17 May 2017 amending Directive 2007/36/EC as regards the encouragement of long-term shareholder engagement (Text with EEA relevance) (https://eur-lex.europa.eu/legal-content/EN/TXT/?uri=CELEX%3A32017L1132).
25. COM (2022), Proposal for a Directive of the European Parliament and of the Council on Corporate Sustainability Due Diligence and amending Directive (EU) 2019/1937. 1_1_183885_prop_dir_susta_en.pdf (https://ec.europa.eu/info/sites/default/files/1_1_183885_prop_dir_susta_en.pdf).
26. Ibid, pg. 21 of the explanatory memorandum.
27. Ibid, pg. 31 of the explanatory memorandum.
28. Interestingly, according to the proposal's explanatory memorandum, companies whose business partner is an SME are also "encouraged to support them in fulfilling the due diligence requirements, in case such requirements would jeopardize the viability of the SME".
29. Ibid, art. 25.

References

Clarkson P. M., Li, Y., Richardson, G. D., & Vasvari, F. P. (2008). Revisiting the relation between environmental performance and environmental disclosure: An empirical analysis. *Accounting, Organizations and Society, 33*(4–5).

Endrikat, J., de Villiers, C., Guenther, T. W., & Guenther, E. M. (2020). Board characteristics and corporate social responsibility: A meta-analytic investigation. *Business & Society*.

EY. (2019). Study on directors' duties and sustainable corporate governance, Available at: Publications Office of the EU https://op.europa.eu/en/publication-detail/-/publication/e47928a2-d20b-11ea-adf7-01aa75ed71a1/language-en, pg. 7.

HLEG. (2018). Final report of the High-Level Expert Group on sustainable finance, available at: European Commission https://ec.europa.eu/info/files/180131-sustainable-finance-final-report_en

ILO. (2017). Tripartite Declaration of Principles concerning Multinational Enterprises and Social Policy (MNE Declaration)—5th Edition (2017), available at: (ENTERPRISES) https://www.ilo.org/empent/areas/mne-declaration/lang--en/index.htm

Lopez de Silanes, F., McCahery, J. A., & Pudschedl, P/C. (2019). ESG performance and disclosure: A cross-country analysis (European Corporate Governance Institute—Law Working Paper 481).

OECD. (2011). *Guidelines for multinational enterprises*, available at: Guidelines—Organisation for Economic Co-operation and Development http://mneguidelines.oecd.org/guidelines/

UN. (2011). *Guiding principles on business and human rights*, available at guidingprinciplesbusinesshr_en.pdf https://digitallibrary.un.org/record/720245?ln=en

CHAPTER 10

Transparency and Standardization of Financial Products Information

Sara Lovisolo

Introduction

The EU Action Plan on Sustainable Finance was designed so as to drive a shift in capital allocation towards sustainable assets only through the lever of transparency (see also Chapter 8). Otherwise stated, the policy measures that make up the sustainable finance package do not prescriptively outline what investment choices should be made by EU financial markets participants, but only rely on the availability of information around those investment choices to achieve the investment shift. The key underlying assumption is that once information around sustainability risks and opportunities has been made available, all financial system actors will inevitably factor in and price those risks and opportunities thus correcting a market failure (see Chapter 3 for discussion on the role of information as a driver of sustainable finance).[1]

S. Lovisolo (✉)
Borsa Italiana, Milano, Italy
e-mail: Sara.Lovisolo@borsaitaliana.it

N. Linciano et al. (eds.), *Information as a Driver of Sustainable Finance*, Palgrave Studies in Impact Finance,
https://doi.org/10.1007/978-3-030-93768-3_10

The Action Plan takes a two-step approach to making sustainability information work. First it creates regulatory use-cases for sustainability information, that is compliance cases. The most important use-cases are the introduction of ESG considerations into investment advice ("suitability test" under MiFID II and Insurance Distribution Directive) and the requirement for all "financial markets participants" (broadly, institutional investors) to report on the adverse impacts of their investment and integrate ESG considerations in their governance, investment policy and risk management under the Sustainable Finance Disclosure Regulation or SFDR. Then it creates tools to make the sustainability input information available and reliable. The tools developed (or in the process of being developed at the time of writing as discussed in Chapter 8) to cater for the information needs created by the above use-cases are:

1. ESG disclosures for benchmarks under the Benchmark Regulation (BMR);
2. The EU Climate Benchmarks under the BMR;
3. The EU Green Bond Standard;
4. The EU Ecolabel for financial products;
5. The product disclosure for sustainable products and products with environmental and social characteristics (SFDR articles 6, 8 and 9).

These can be matched to the use-cases as in Table 10.1.

Table 10.1 Use-cases and tools developed under the EU Action Plan

Use-cases	*Disclosure tools*	*Input data and methodologies*
Institutional investors' ESG obligations	• ESG disclosures for benchmarks • EU Climate benchmarks • EU Green Bond Standard	EU Taxonomy NFRD disclosures (to be replaced by CSRD disclosures)
ESG in suitability test (retail investors)	• Product disclosures for ESG funds • EU Green Bond Standard • EU Ecolabel for financial products • EU Climate benchmarks	EU Taxonomy NFRD disclosures (to be replaced by CSRD disclosures)

The EU Taxonomy (discussed in Chapter 8) and the Non-financial Reporting Directive (in perspective the Corporate Sustainability Reporting Directive; discussed in Chapters 4 and 9) are fundamental building blocks of the EU sustainable finance regulatory architecture and feed into the above tools.

It must be noted that the 2 use-cases and the 5 tools explored here are only those included in the EU Action Plan on Sustainable Finance, and should be examined in the context of the wider EU sustainability reporting environment, which includes other measures, such as the Second Shareholder Rights Directive,[2] the proposed "pillar 3" disclosures under Solvency II and the Capital Requirements Regulations 2, sustainability considerations included in IORP II and InvestEU.[3]

In this chapter, we'll explore the objectives, fitness for purpose and limits of the tools that support the retail use-case, in order to identify corrections measures that might be addressed in future iterations of the EU sustainable finance strategy.[4]

The Retail Use-Case for Sustainability Information

The EU Sustainable Finance Retail Strategy as Part of the EU Action Plan

One of the recommendations of the High-Level Expert Group on Sustainable Finance (HLEG) concerned the development of an EU retail sustainable finance strategy, based on the observation that "European household savings represent over 40% of total financial assets in the EU". The recommended strategy was based on three pillars: (1) the introduction of sustainability-related considerations in investment advice; (2) the creation of an EU label for green investment products and (3) the introduction of rules for the attribution of fund names that make reference to sustainability. These measures were justified on the basis that "EU citizens expect sustainable funds to enable them to have a positive impact on the economy; but they lack the concrete tools to identify corresponding investment products".

All the above recommendations have translated into corresponding actions in the EU Action Plan. As already discussed in Chapter 8, the introduction of sustainability consideration into investment advice has been captured by Action 4 and the creation of an EU label for green investment products has translated into Action 2.3 ("*The Commission will explore the use of the EU Ecolabel framework for certain financial products, to be applied once the EU sustainability taxonomy is adopted*"). The recommendation concerning the use of sustainability references in investment fund names has instead been upgraded to encompass wider ranging disclosures of sustainable investment products and products "promoting ESG characteristics" under the SFDR, as we'll see below.

In the proposed reading of the EU retail sustainable finance strategy—as suggested by the HLEG and implemented by the Commission in the Action Plan—the first measure, the introduction of sustainability considerations in investment advice, is set apart from the remaining two as the latter are ancillary to the former. The label and the information associated with investment products are means to an end. The end—or use-case—is the ability of retail investors to express preferences for the integration of sustainability considerations into their investment choices, which implicitly requires that investment advisors are enabled to respond to those needs by using reliable signalling and information tools (Fig. 10.1).

As discussed further below, the initial choice of tools—SFDR product disclosures and green label—can be expanded to include the EU Green Bond Standard, and also the official EU methodologies for the development of climate benchmarks under the Benchmark Regulation (respectively action 2.1 and 5 of the EU Action Plan).

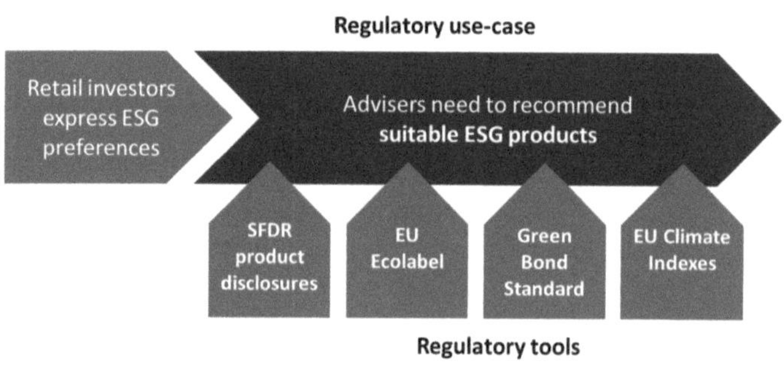

Fig. 10.1 Retail investor use-case

The Delegated Regulation to MiFID II on the Integration of ESG Considerations in the Suitability Test

Action 4 of the EU Action Plan states: "Subject to the outcome of its impact assessment, the Commission will amend the MiFID II and IDD delegated acts in Q2 2018 to ensure that sustainability preferences are taken into account in the suitability assessment. Based on these delegated acts, the Commission will invite the European Securities Markets Authority (ESMA) to include provisions on sustainability preferences in its guidelines on the suitability assessment to be updated by Q4 2018".[5]

In August 2021, the European Commission published in the Official Journal the following Delegated Regulation (EU) 2021/1253, amending a delegated act to MiFID II, designed to ensure insurance distributors and investment firms take environmental, social and governance ("ESG") issues into account when advising customers.[6]

The Delegated Regulation has been produced under Articles 24(13) and 25(8) of the MiFID II Directive, and will apply from 2 August 2022.

In June 2020, the European Commission had published draft delegated acts on suitability assessments that would amend existing Level 2 measures under MiFID2. The consultation period for the proposed delegated act closed on 6 July 2020.

The main thrust of the new regulation is that "investment firms that provide investment advice and portfolio management services should clearly explain the distinction between financial products that promote environmental or social characteristics and financial products that pursue sustainable investment objectives".

The new delegated act provides the following definition: (7) "sustainability preferences" means a client's or potential client's choice as to whether and, if so, to what extent, one or more of the following financial instruments shall be integrated into his or her investment:

(a) a financial instrument for which the client or potential client determines that a minimum proportion shall be invested in environmentally sustainable investments as defined in Article 2, point (1), of Regulation (EU) 2020/852 of the European Parliament and of the Council;
(b) a financial instrument for which the client or potential client determines that a minimum proportion shall be invested in sustainable investments as defined in Article 2, point (17), of Regulation (EU) 2019/2088 of the European Parliament and of the Council;

(c) a financial instrument that considers principal adverse impacts on sustainability factors where qualitative or quantitative elements demonstrating that consideration are determined by the client or potential client. Fundamentally, in providing investment advice, banks and investment advisors will have to determine with their clients whether they have a minimum share of the instruments that addresses either environmental objectives as determined by the EU Taxonomy Regulation (see point (a) above) or a minimum share or sustainable investment objectives more broadly, as defined under the SFDR (point b above). In addition, the client can clarify quantitative elements that determine that certain ESG factors of the client's choice are taken into account by the instrument. In practice, the investment advisor and the client are expected to agree two percentages—which could also be zero—regarding the share of investment in (i) Taxonomy-aligned activities and (ii) sustainable investment as defined under the SFDR, instrument by instrument. Alternatively, they should agree which ESG factors and adverse impacts the clients want to control, and what quantitative indicators would satisfy the client that his concerns have been met.

Implications of the Delegated Regulation

The first observation regarding the delegated regulation summarized in the previous section, is that its object is "financial instruments". In EU Regulation, these are defined in Section C of MiFID II Directive by way of the below positive list:

(1) Transferable securities;
(2) Money-market instruments;
(3) Units in collective investment undertakings;
(4) Options, futures, swaps, forward rate agreements and any other derivative contracts relating to securities, currencies, interest rates or yields, emission allowances or other derivatives instruments, financial indices or financial measures which may be settled physically or in cash;
(5) Options, futures, swaps, forwards and any other derivative contracts relating to commodities that must be settled in cash or may be settled in cash at the option of one of the parties other than by reason of default or other termination event;

(6) Options, futures, swaps, and any other derivative contract relating to commodities that can be physically settled provided that they are traded on a regulated market, a MTF or an OTF, except for wholesale energy products traded on an OTF that must be physically settled;
(7) Options, futures, swaps, forwards and any other derivative contracts relating to commodities, that can be physically settled not otherwise mentioned in point 6 of this Section and not being for commercial purposes, which have the characteristics of other derivative financial instruments;
(8) Derivative instruments for the transfer of credit risk;
(9) Financial contracts for differences;
(10) Options, futures, swaps, forward rate agreements and any other derivative contracts relating to climatic variables, freight rates or inflation rates or other official economic statistics that must be settled in cash or may be settled in cash at the option of one of the parties other than by reason of default or other termination event, as well as any other derivative contracts relating to assets, rights, obligations, indices and measures not otherwise mentioned in this Section, which have the characteristics of other derivative financial instruments, having regard to whether, inter alia, they are traded on a regulated market, OTF or an MTF;
(11) Emission allowances consisting of any units recognized for compliance with the requirements of Directive 2003/87/EC (Emissions Trading Scheme).

Even if we exclude cases (5) to (10) from the scope of customary application of retail investment advice, the scope of the draft regulation does not match the scope of the SFDR, which defines[7] "financial products" as

(a) a portfolio managed in accordance with point (6) of this Article;
(b) an alternative investment fund (AIF);
(c) an IBIP;
(d) a pension product;
(e) a pension scheme;
(f) a UCITS;
or (g) a PEPP.

It should not be given for granted that in practice investment advice under MiFID II and IDD will coincide with the scope of the SFDR, given the diversity of retail investment approaches across the 27 Member States of the EU. For example, it can be very common for investment advisers to put forward sovereign bonds as a low-risk investment proposal, while retail investors will very often inquire about the possibility to invest in single stocks or bonds (as opposed to investment funds). For a breakdown of the assets held by households in the Eurozone see the following Figure which highlights how equities and bonds—held directly and not as part of mutual fund portfolios—constitute a significant proportion of total holdings (Fig. 10.2).

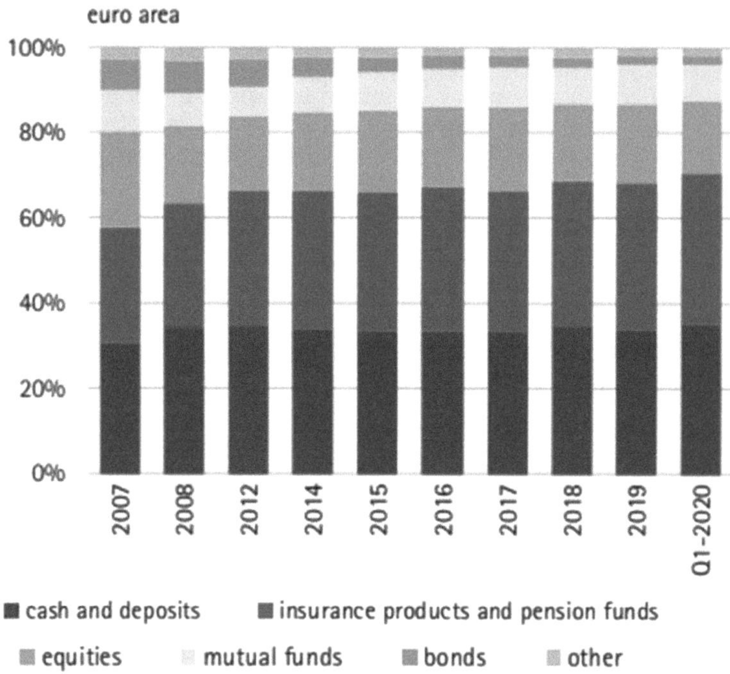

Fig. 10.2 Breakdown of household financial assets (*Source* CONSOB, 2020)

This raises the questions of how an investment adviser should assess whether a transferable security or a money-market instrument (for instance) pursues environmental or social objectives, or promotes environmental characteristics—as the guidance provided in SFDR at Level 1 or Level 2 does not cover single instruments.

The second observation is that the draft regulation is totally reliant on the definitions of sustainable investment and investment with ESG characteristics introduced by the SFDR which had portfolios in mind, and where the agent that "pursues" (environmental or social objective) or "promotes" (ESG characteristics) is implicitly the investment strategy put in place by the investment firm (or "financial market participant" in SFDR parlance).

The extension of SFDR definitions to government bonds, single listed bonds or shares, or options becomes problematic as the same "agency" (the fact of pursuing or promoting) cannot be attributed as easily to a government or a company (the issuers of government bonds, or corporate bonds or equity shares).

We'll discuss later in detail what options are available to address this gap. An initial overview of the toolbox (both regulatory and market based) is provided in the Table 10.2.

Table 10.2 Overview of tools supporting ESG investment advice

Financial instruments	*EU Action Plan measures*	*Other market or regulatory tools*
Transferable securities (bonds)	EU Green Bond standard	ESG in credit ratings
Transferable securities (equity shares)	–	ESG ratings
Money markets instruments	–	ESG in credit ratings
Units in collective investment undertakings	SFDR product disclosures EU Ecolabel BMR ESG disclosures EU Climate Benchmarks	Fund ratings
Options, futures, swaps relating to securities, financial indices and currencies, commodities	BMR ESG disclosures (for index derivatives) EU Climate Benchmarks	

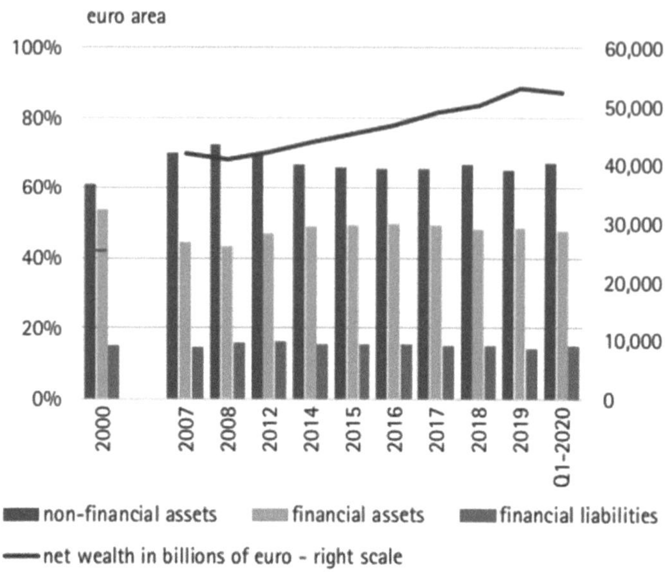

Fig. 10.3 Household net wealth in the Eurozone (*Source* CONSOB, 2020)

The impact of the measures introduced by Action 4 cannot however be overstated, either in terms of its potential to mobilize savings towards sustainable assets or the liability risks it raises for investment advisers. As shown by the following Figure, only in the Eurozone the net wealth of households has been steadily growing since the Great Financial Crisis of 2008, exceeding €50tn. With financial assets amounting to over €30tn in 2020, the potential impact of integrating ESG consideration in household decisions is considerable (Fig. 10.3).

Overview of the EU Toolbox Supporting the Retail Use-Case

Standardization of Product Disclosures Under the SFDR

The SFDR introduces transparency requirements at the product level, both as part of pre-contractual information and of periodic reporting associated with investment products.

As already mentioned in Chapter 8, Article 6 states that financial market participants and financial advisers "*shall include descriptions of the following in pre-contractual disclosures: (a) the manner in which sustainability risks are integrated into their investment decisions; and (b) the results of the assessment of the likely impacts of sustainability risks on the returns of the financial products they make available*".

This obligation applies to all the products made available by asset managers who in principle take account of sustainability factors in their investment approach as per Article 4. That is, if an asset manager does not resort to the "explain" option open to smaller entities (those with fewer than 500 employees), the firm will have to include the above descriptions in the pre-contractual documents of all its products, whether marketed as sustainable (in broad terms) or not.

Article 7 goes on to specify that the pre-contractual information mentioned in Article 6 should include "(a) a clear and reasoned explanation of whether, and, if so, how a financial product considers principal adverse impacts on sustainability factors; (b) a statement that information on principal adverse impacts on sustainability factors is available" in the periodic reports. This Article creates a direct link between the Principal Adverse Impact statement drawn up at the entity level (see previous Chapter 7 for more details) and all the products "manufactured" by the financial firm, the idea being that the overall exposure of an asset manager or a pension fund to sustainability risks should be the sum of the exposure of the single investment products the financial market participant manages.

For products that are not marketed as "sustainable" or "with ESG characteristics", no further standardization is envisaged by the level 1 regulation. Hence, for the disclosure of principal adverse impacts at the product level in both pre-contractual information and periodic reports, financial markets participants are expected to make reference to the indicators recommended by the ESAs in their proposal for the Regulatory Technical Standards (RTS) associated with the SFDR. A list of indicators—as put forward for consultation by the ESAs[8]—is provided in Tables 10.3 and 10.4.

The approach adopted by the regulator is to provide indicators that can be compared—like for like—across investment products. Theoretically, an investment adviser or a retail investor could pick two investment products and compare, for instance, the weighted average carbon intensity (indicator 3 in Table 10.3) and draw conclusions on the relative environmental

Table 10.3 Proposed disclosures for the Principal Adverse Impact statement (draft RTS)—mandatory indicators

#	*Disclosure area*	*Adverse sustainability indicator*	*Metric*
1	Climate and other environment-related indicators/Greenhouse Gas Emissions	Carbon emissions (broken down by Scope 1, 2, 3 carbon emissions—including agriculture, forestry and other land use [AFOLU] emissions and in total)	(f) 'total carbon emissions' shall be calculated in accordance with the following formula $\sum_{n}^{i}\left(\frac{current\ value\ of\ investment_i}{investee\ company's\ enterprise\ value_i} \times investee\ company's\ Scope\ 1,\ 2\ and\ 3\ carbon\ emissions_i\right)$
2	Climate and other environment-related indicators/Greenhouse Gas Emissions	Carbon footprint	(g) 'carbon footprint' shall be calculated in accordance with the following formula $\frac{\sum_{n}^{i}\left(\frac{current\ value\ of\ investment_i}{investee\ company's\ enterprise\ value_i} \times investee\ company's\ Scope\ 1,\ 2\ and\ 3\ carbon\ emissions_i\right)}{current\ value\ of\ all\ investments (€M)}$
3	Climate and other environment-related indicators/Greenhouse Gas Emissions	Weighted average carbon intensity	(i) 'carbon intensity' shall be calculated in accordance with the following formula $\sum_{n}^{i}\left(\frac{current\ value\ of\ investment_i}{current\ value\ of\ all\ invertments\ (€M)} \times \frac{investee\ company's\ scope\ 1,\ 2\ and\ 3\ carbon\ emissions_i}{investee\ company's\ €M\ revenue_i}\right)$
4	Climate and other environment-related indicators/Greenhouse Gas Emissions	Solid fossil fuel sector exposure	Share of investment in solid fossil fuel sectors
5	Climate and other environment-related indicators/Energy performance	Total energy consumption from non-renewable energy sources and share of non-renewable energy consumption	1. Total energy consumption of investee companies from non-renewable energy sources (in GWh) expressed as a weighted average 2. Shared on non-renewable energy consumption of investee companies from non-renewable energy sources compared to renewable energy sources, expressed as a percentage

#	*Disclosure area*	*Adverse sustainability indicator*	*Metric*
6	Climate and other environment-related indicators/Energy performance	Breakdown of energy consumption by type of non-renewable sources of energy	Share of energy from non-renewable sources used by investee companies broken down by each non-renewable energy source
7	Climate and other environment-related indicators/Energy performance	Energy consumption intensity	Energy consumption of investee companies per million EUR of revenue of those companies (in GWh), expressed as a weighted average
8	Climate and other environment-related indicators/Energy performance	Energy consumption intensity per sector	Energy consumption intensity per million EUR of revenue of investee companies, per NACE sector (in GWh), expressed as a weighted average
9	Climate and other environment-related indicators/Biodiversity	Biodiversity and ecosystem preservation practices	1. Share of all investments in investee companies that do not assess, monitor or control the pressures corresponding to the indirect and direct drivers of biodiversity and ecosystem change 2. Share of all investee companies that do not assess, monitor or control the pressures corresponding to the indirect and direct drivers of biodiversity and ecosystem change
10	Climate and other environment-related indicators/Biodiversity	Natural species and protected areas	1. Share of investments invested in investee companies whose operations affect IUCN Red List species and/or national conservation list species; 2. Share of investments in investee companies with operational sites owned, leased, managed in, or adjacent to, protected areas and areas of high biodiversity value outside protected areas

(continued)

Table 10.3 (continued)

#	*Disclosure area*	*Adverse sustainability indicator*	*Metric*
11	Climate and other environment-related indicators/Biodiversity	Deforestation	1. Share of investments in entities without a deforestation policy 2. Share of investee companies without a deforestation policy
12	Climate and other environment-related indicators/Water	Water emissions	Weight in tonnes of water emissions generated by investee companies per million EUR invested, expressed as a weighted average
13	Climate and other environment-related indicators/Water	Exposure to areas of high-water stress	1. Share of investments in investee companies with sites located in areas of high-water stress 2. Share of investee companies with sites located in areas of high-water stress
14	Climate and other environment-related indicators/Water	Untreated discharged wastewater	Total amount in cubic meters of untreated wastewater discharged by the investee companies expressed as a weighted average
15	Climate and other environment-related indicators/Waste	Hazardous waste ratio	Weight in tonnes of hazardous waste generated by investee companies per million EUR invested, expressed as a weighted average
16	Climate and other environment-related indicators/Waste	Non-recycled waste ratio	Weight in tonnes of non-recycled waste generated by investee companies per million EUR invested, expressed as a weighted average
17	Social and employee matters	Implementation of fundamental ILO Conventions	1. Share of investments in entities without due diligence policies on issues addressed by the fundamental ILO Conventions 1–8 2. Share of investee companies without due diligence policies on issues addressed by the fundamental ILO Conventions 1 to 8
18	Social and employee matters	Gender pay ga	Average gender pay gap of investee companies

#	*Disclosure area*	*Adverse sustainability indicator*	*Metric*
19	Social and employee matters	Excessive CEO pay ratio	Average ratio within investee companies of the annual total compensation for the highest compensated individual to the median annual total compensation for all employees (excluding the highest-compensated individual)
20	Social and employee matters	Board gender diversity	Average ratio of female to male board members in investee companies
21	Social and employee matters	Insufficient whistleblower protection	1. Share of investments in entities without policies on the protection of whistleblowers 2. Share of investee companies without policies on the protection of whistleblowers
22	Social and employee matters	Investment in investee companies without workplace accident prevention policies	1. Share of investments in investee companies without a workplace accident prevention policy 2. Share of investee companies without a workplace accident prevention policy
23	Human rights	Human rights policy	1. Share of investments in entities without a human rights policy 2. Share of investee companies without a human rights policy
24	Human rights	Due diligence	1. Share of investments in entities without a due diligence process to identify, prevent, mitigate and address adverse human rights impacts 2. Share of investee companies without a due diligence process to identify, prevent, mitigate and address adverse human rights impacts

(continued)

Table 10.3 (continued)

#	*Disclosure area*	*Adverse sustainability indicator*	*Metric*
25	Human rights	Processes and measures for preventing trafficking in human beings	1. Share of investments in investee companies/all investments without policies against trafficking in human beings 2. Share of all investments exposed to entities without international framework agreements combating trafficking in human beings
26	Human rights	Operations and suppliers at significant risk of incidents of child labour	1. Share of the investments in investee companies /share of all investments exposed to operations and suppliers at significant risk of incidents of child labour exposed to hazardous work in terms of geographic areas or type of operation 2. Share of investee companies/share of all investments exposed to operations and suppliers at significant risk of incidents of forced or compulsory labour in terms in terms of geographic areas and/or the type of operation
27	Human rights	Operations and suppliers at significant risk of incidents of forced or compulsory labour	Share of the investments in investee companies/share of all investments exposed to operations and suppliers at significant risk of incidents of forced or compulsory labour in terms in terms of geographic areas and/or the type of operation
28	Human rights	Number and nature of identified cases of severe human rights issues and incidents	Number and nature of cases of severe human rights issues and incidents connected to investee companies
29	Human rights	Exposure to controversial weapons (land mines and cluster bombs)	Any investment in entities involved in the manufacture or selling of controversial weapons (land mines and cluster bombs)

#	*Disclosure area*	*Adverse sustainability indicator*	*Metric*
30	Anti-corruption and anti-bribery	Anti-corruption and anti-bribery policies	1. Share of investments in entities without policies on anti-corruption and anti-bribery consistent with the United Nations Convention against Corruption 2. Share of investee companies without policies on anti-corruption and bribery consistent with the United Nations Convention against Corruption
31	Anti-corruption and anti-bribery	Cases of insufficient action taken to address breaches of standards of anti-corruption and anti-bribery	1. Share of investments in investee companies with identified insufficiencies in actions taken to address breaches in procedures and standards of anti-corruption and anti-bribery 2. Share of investee companies with insufficiencies in actions taken to address breaches in procedures and standards of anti-corruption and anti-bribery
32	Anti-corruption and anti-bribery	Number of convictions and amount of fines for violation of anti-corruption and anti-bribery laws	Numbers of convictions and amount of fines for violations of anti-corruption and anti-bribery laws by investee companies

performance of the two products, opting for the product with a lower weighted average carbon intensity. This example raises already significant issues. It's not to be given for granted that a product with lower carbon intensity has better environmental attributes, as a lower carbon intensity can just be the result of exposure to industries with no contribution to the transition to a low-carbon economy (such as media, pharmaceuticals, insurance, financials). Hence it's likely to be expected that the adverse impact indicators per se will require commentary—either on the part of the product manufacturer or the investment adviser—to translate into meaningful and decision-useful information for retail investors.

However, the most relevant disclosures are those associated with articles 8 and 9 of the SFDR. Article 8 is concerned with—and at the same time introduces in the regulatory environment—financial products that promote "among other characteristics, environmental or social characteristics, or a combination of those characteristics, provided that the companies in which the investments are made follow good governance practices". This definition is a remarkable innovation, as it creates an official EU definition of ESG products that cuts across the classic definitions built over the years by market practice which identify investment strategies that asset managers and investors more broadly have learnt to translate into portfolio management approaches. These are: exclusionary strategies, best-in-class, norms-based screening, engagement, ESG integration, thematic investment and impact investment. A clear limit of this definition however is that it assumes that investments are always made in companies—thus excluding other assets such as real estate, commodities or government bonds.

For products falling under Article 8, pre-contractual documents "shall include (a) information on how those characteristics are met; (b) if an index has been designated as a reference benchmark, information on whether and how this index is consistent with those characteristics". The second requirement places significant importance—and creates a fundamental use-case—for another regulatory initiative of the EU Action Plan, the introduction of ESG disclosures for benchmarks and the creation of official EU methodologies for climate benchmarks, to be discussed in the next section.

Article 9 is even more reliant on indexes as benchmarks than Article 8. It states that "where a financial product has sustainable investment as its objective and an index has been designated as a reference benchmark, the information to be disclosed (…) shall be accompanied by the following: (a) information on how the designated index is aligned with that objective; (b) an explanation as to why and how the designated index aligned with that objective differs from a broad market index". It also clarifies that "where a financial product has a reduction in carbon emissions as its objective, the information to be disclosed (...) shall include the objective of low carbon emission exposure in view of achieving the long-term global warming objectives of the Paris Agreement". Here the SFDR creates a further category of ESG products—those with sustainable investment as their objectives, where sustainable investment is defined in Article 1 as "an investment in an economic activity that contributes to an environmental objective, as measured, for example, by key resource efficiency indicators on the use of energy, renewable energy, raw materials, water and land, on the production of waste, and greenhouse gas emissions, or on its impact on biodiversity and the circular economy, or an investment in an economic activity that contributes to a social objective, in particular an investment that contributes to tackling inequality or that fosters social cohesion, social integration and labour relations, or an investment in human capital or economically or socially disadvantaged communities, provided that such investments do not significantly harm any of those objectives and that the investee companies follow good governance practices, in particular with respect to sound management structures, employee relations, remuneration of staff and tax compliance". This definition seems to restrict sustainable investment to the impact strategy, only one of the seven typical ESG investment strategies mentioned above.

Since for sustainable investment products under Article 9 disclosures are either associated with explanations regarding the index used as benchmarks or alignment to a temperature scenario in the context of the Paris Agreement on climate, a review of Action 5 is provided in the next section.

ESG Disclosures for Benchmarks Under the Benchmarks Regulation

The final report by the ESAs on the RTS supporting the SFDR[9] states that "it is appropriate to standardise the metrics used to assess certain adverse impacts which are considered to be measurable and important to provide a common reference point for the purposes of identifying which of those impacts are principal and to ensure comparability. To ensure coherence between other sustainability-related disclosures, where relevant, those common reference indicators should be based on similar indicators used in the minimum standards for the EU Climate Transition Benchmarks and EU Paris-aligned Benchmarks supplementing Regulation (EU) 2016/1011". Many of the indicators listed in Tables 10.3 and 10.4 are indeed drawn from the indicators set out by the Benchmarks Regulation.

By way of background, in compliance with Action 4 of the EU Action Plan, in early 2019 the European Parliament and the Council reached a political agreement on a regulation amending Regulation (EU) 2016/1011, also known as the Benchmark Regulation or BMR. This amendment introduced methodologies for the development of two climate benchmarks already mentioned in Chapter 6—the EU Paris-aligned Benchmark and the EU Climate Transition Benchmarks. It also introduced sustainability-related disclosure requirements for all benchmarks—or reference indexes—published by benchmark administrators registered or recognized in the European Union. This was a significant innovation in the global disclosure landscape, with impacts that go beyond the boundaries of the European Union—given the international reach of most index providers. A new important use-case for disclosures under the NFRD has been established, which requires careful consideration of its vertical alignment with both corporate disclosures and the institutional use of sustainability data by investors under the SFDR. The new disclosure requirements came into force by the end of 2020.

The final regulation, Regulation (EU) 2019/2089, states[10] that "*users of benchmarks do not always have the necessary information on the extent to which the methodology of the benchmark administrators takes into account ESG factors. Such information is often scattered or non-existent, and does not allow for an effective comparison across borders for investment purposes. To enable market participants to make well-informed choices, all benchmark administrators, with the exception of administrators of interest rate and foreign exchange benchmarks, should be required to disclose in their benchmark statement whether or not their benchmarks or families of benchmarks pursue ESG objectives, and whether or not the benchmark administrator offers such benchmarks*". Hence the provision in the delegated acts of the regulation of a list of very detailed disclosure requirements, fundamentally applicable to ESG benchmarks.[11]

There are two types of sustainability disclosure requirements for benchmarks:

Table 10.4 Proposed disclosures for the Principal Adverse Impact statement (draft RTS)—optional indicators

#	*Disclosure area*	*Adverse sustainability indicator*	*Metric*
33	Emissions	Emissions of inorganic pollutants	Tonnes of inorganic pollutants equivalent per million EUR invested
34	Emissions	Emissions of air pollutants	
35	Emissions	Emissions of ozone depletion substances	Tonnes of ozone depletion substances equivalent per million EUR invested
36	Emissions	Investing in companies without carbon emission reduction initiatives	1. Share of investments in investee companies without carbon emission reduction initiatives 2. Share of investee companies without carbon emission reduction initiatives
37	Water, waste and material	Water usage: Total amount of water consumed and reclaimed, broken down per sector where relevant	Average amount of water consumed and reclaimed by the investee companies (in cubic meter) per million EUR of revenue of investee companies
38	Water, waste and material	Water recycled and reused	Weighted average percentage of water recycled and reused by investee companies
39	Water, waste and material	Investing in companies without water management initiatives	1. Share of investments in investee companies with no water management initiatives 2. Share of investee companies with no water management initiatives
40	Water, waste and material	Land degradation, desertification, soil sealing	Land degradation, desertification and soil sealing as a percentage of land by the investee companies FMP invests in

(continued)

Table 10.4 (continued)

#	*Disclosure area*	*Adverse sustainability indicator*	*Metric*
41	Water, waste and material	Investing in companies without sustainable land/forestry/agriculture practices	1. Share of investments in investee companies without sustainable land/forestry/agriculture practices policies 2. Share of investee companies without sustainable land/forestry/ agriculture practices policies
42	Water, waste and material	Investing in companies without sustainable oceans/seas practices	1. Share of investments in investee companies without sustainable oceans/seas practices policies 2. Share of investee companies without sustainable oceans/seas practices policies
43	Green securities	Share of securities not certified as gree	Share of securities in investments not certified as green
44	Social and employee matters	Number/rate of accidents, injuries, fatalities, frequency	Number/rate of accidents, injuries, fatalities frequency in investee companies
45	Social and employee matters	Number of days lost for injuries, accidents, fatalities, illness	Number of workdays lost to injuries, accidents, fatalities, illness of investee companies

#	*Disclosure area*	*Adverse sustainability indicator*	*Metric*
46	Social and employee matters	Supplier code of conduct	1. Share of investments in investee companies without any supplier code of conduct (against unsafe working conditions, precarious work, child labour and forced labour) 2. Percentage of specific control and/or certified compliance for this code of conduct among suppliers of investee companies
48	Social and employee matters	Grievance/complaints handling mechanism	1. Share of investments in investee companies without any grievance/complaints handling mechanism 2. Share of investee companies without any grievance/complaints handling mechanism
49	Social and employee matters	Incidents of discrimination	1. Number of incidents of discrimination reported in investee companies 2. Number of incidents of discrimination leading to sanctions in investee companies
50	Social and employee matters	Lack of separation of CEO and Chair functions on the boards of investee companies	1. Share of investments in investee companies without separate CEO and Chair functions 2. Share of investee companies without separate CEO and Chair functions
51	Social securities	Share of securities not certified as social	Share of securities in investments not certified as social

(a) ESG disclosures: disclosure of ESG indicators in the benchmark statement at the individual benchmark of benchmark family level. A list of 78 indicators and the associated disclosure template has been provided in the delegated acts.[12] These indicators are associated with the relevant asset class of the benchmark: equity, corporate bonds, sovereign debt, commodity and other. Disclosures are mandatory for ESG benchmarks belonging to any asset class and voluntary for other benchmarks. In their disclosure, benchmark administrators are also required to provide details on the nature of the data used—whether proprietary or from other sources, and whether reported or estimated. They also have to reference the standards used for the calculation of the indicators.
(b) Paris-alignment: the most innovative disclosure is the explanation of the benchmark alignment with the decarbonization objectives of the Paris Agreement. The delegated acts provide a template for these disclosures. The benchmark administrator has to (1) declare whether the benchmark align with the target of reducing carbon emissions or the attainment of the objectives of the Paris Agreement; (2) disclose the temperature scenario, in accordance with international standards, used for the alignment with the target of reducing GHG emissions or attaining of the objectives of the Paris Agreement, the name of the scenario provider and a link to the scenario; (3) the methodology used for the measurement of the alignment with the temperature scenario. These disclosures apply to all significant equity and bond benchmark (i.e. indexes with more than €50bn benchmarked to them) from 2020, and from 2022 to all benchmarks.

Sustainability-related benchmark disclosures are set out by asset class. The asset classes covered are: equity, fixed income, sovereign debt, commodities and other (private equity and debt, real estate).

A list of indicators is provided and must be adhered to, without the ability of the index administrator to choose the most material indicators against which to disclose. Some of the indicators are set as voluntary, in consideration of data availability issues. This aligns with the draft RTS of the SFDR, which provides two separate lists of indicators, a mandatory list and a list the financial market participants can choose from.

Benchmark administrators also have to disclose:

(a) Description of data sources used to provide information on the ESG factors in the benchmark statement, by describing how the data used to provide information on the ESG factors in the benchmark statement are sourced and whether, and to what extent, data are estimated or reported.
(b) Reference standards, by listing the supporting standards used for calculating the indicators.

The indicators are described in Tables 10.5, 10.6, 10.7, 10.8, 10.9, and 10.10.

These disclosures are an important tool complementing the disclosures under Articles 8 and 9 of the SFDR, insofar as they provide methodologies and input data that can be used by asset managers to explain the alignment of a fund with a temperature scenario in the context of the Paris Agreement (see Table 10.5) or the characteristics of the ESG index used to benchmark the performance of a fund pursuant to Article 8 and Article 9.

However, as the sophistication of the disclosure shows, this input data—very useful for asset managers in discharging their obligations under Articles 8 and 9 of the SFDR—seems of limited help with regard to the needs of investment advisers and retail investors when considering ESG investment choice. Again, we can take the case of the greenhouse gas intensity of a benchmark (see indicator 12 in Table 10.6)—along the same line of argument followed when discussing the SFDR. While it's true that this is a very standardized and comparable type of disclosure, its value for decision purpose is unclear unless it's part of a broader framework that links such disclosure to real-world impact (Tables 10.7, 10.8, 10.9, and 10.10).

EU Climate Benchmarks

The regulation amending the Benchmark Regulation described in the previous section[13] also introduced methodologies for two EU climate benchmarks, the EU Paris-aligned Benchmark and the EU Climate Transition Benchmarks. The rationale for this regulatory intervention in a field so far left to market competition and initiative—the development of indexes incorporating climate-related considerations—is justified by the policy maker in the impact assessment on the basis of lack of reliable methodologies according to feedback from stakeholders, and to

Table 10.5 Paris alignment—applicable to all equity and bond benchmarks

#	*Disclosure area*	*Indicator*	*Metric*
1	Paris alignment	Does the benchmark align with the target of reducing carbon emissions or the attainment of the objectives of the Paris Agreement?	Yes/No
2	Paris alignment	The temperature scenario, in accordance with international standards, used for the alignment with the target of reducing GHG emissions or attaining of the objectives of the Paris Agreement	Narrative
3	Paris alignment	The name of the provider of the temperature scenario used for the alignment with the target of reducing GHG emissions or the attainment of the objectives of the Paris Agreement	Narrative
4	Paris alignment	The methodology used for the measurement of the alignment with the temperature scenario	Narrative
5	Paris alignment	The hyperlink to the website of the temperature scenario used	Object

Table 10.6 ESG disclosures for equity benchmarks

#	*Disclosure area*	*Indicator*	*Mandatory/voluntary disclosure*
6	Combined ESG factors	Weighted average ESG rating of the benchmark	Voluntary
7	Combined ESG factors	Overall ESG rating of top ten benchmark constituents by weighting in the benchmark	Voluntary
8	Environmental	Weighted average environmental rating of the benchmark	Voluntary
9	Environmental	Exposure of the benchmark portfolio to renewable energy as measured by capital expenditures (CapEx) in those activities (as a share of total CapEx by energy companies included in the portfolio)	Voluntary
10	Environmental	Exposure of the benchmark portfolio to climate-related physical risks, measuring the effects of extreme weather events on companies' operations and production or on the different stages of the supply chain (based on issuer exposure)	Voluntary
11	Environmental	Degree of exposure of the portfolio to the sectors listed in Sections A to H and Section L of Annex I to Regulation (EC) No. 1893/2006 of the European Parliament and of the Council2 as a percentage of the total weight in the portfolio	Mandatory
12	Environmental	Greenhouse gas (GHG) intensity of the benchmark	Mandatory

(continued)

Table 10.6 (continued)

#	*Disclosure area*	*Indicator*	*Mandatory/voluntary disclosure*
13	Environmental	Percentage of GHG emissions reported versus estimated	Mandatory
14	Environmental	Exposure of the benchmark portfolio to companies the activities of which fall under Divisions 05 to 09, 19 and 20 of Annex I to Regulation (EC) No. 1893/2006	Mandatory
15	Environmental	Exposure of the benchmark portfolio to activities included in the environmental goods and services sector, as defined in Article 2, point (5) of Regulation (EU) No. 691/2011 of the European Parliament and of the Council	Mandatory
16	Social	Weighted average social rating of the benchmark	Voluntary
17	Social	International treaties and conventions, United Nations principles or, where applicable, national law used in order to determine what constitutes a "controversial weapon"	Mandatory
18	Social	Weighted average percentage of benchmark constituents in the controversial weapons sector	Mandatory
19	Social	Weighted average percentage of benchmark constituents in the tobacco sector	Mandatory

#	*Disclosure area*	*Indicator*	*Mandatory/voluntary disclosure*
20	Social	Number of benchmark constituents subject to social violations (absolute number and relative divided by all benchmark constituents), as referred to in international treaties and conventions, United Nations principles and, where applicable, national law	Mandatory
21	Social	Exposure of the benchmark portfolio to companies without due diligence policies on issues addressed by the fundamental International Labor Organisation Conventions 1 to 8	Mandatory
22	Social	Weighted average gender pay gap	Mandatory
23	Social	Weighted average ratio of female to male board members	Mandatory
24	Social	Weighted average ratio of accidents, injuries, fatalities	Mandatory
25	Social	Numbers of convictions and amount of fines for violations of anti-corruption and anti-bribery laws	Mandatory
26	Governance	Weighted average governance rating of the benchmark	Voluntary
27	Governance	Weighted average percentage of board members who are independent	Mandatory
28	Governance	Weighted average percentage of female board members	Mandatory

Table 10.7 SG disclosures for fixed income benchmarks

#	*Disclosure area*	*Indicator*	*Mandatory/voluntary disclosure*
29	Combined ESG factors	Weighted average ESG rating of the benchmark	Voluntary
30	Combined ESG factors	Overall ESG rating of top ten benchmark constituents by weighting in the benchmark	Voluntary
31	Environmental	Weighted average environmental rating of the benchmark	Voluntary
32	Environmental	Exposure of the benchmark portfolio to renewable energy as measured by capital expenditures (CapEx) in those activities (as a share of total CapEx by energy companies included in the portfolio)	Voluntary
33	Environmental	Exposure of the benchmark portfolio to climate-related physical risks, measuring the effects of extreme weather events on companies' operations and production or on the different stages of the supply chain (based on issuer exposure)	Voluntary
34	Environmental	Degree of exposure of the portfolio to the sectors listed in Sections A to H and Section L of Annex I to Regulation (EC) No. 1893/2006 of the European Parliament and of the Council2 as a percentage of the total weight in the portfolio	Mandatory
35	Environmental	Greenhouse gas (GHG) intensity of the benchmark	Mandatory

#	*Disclosure area*	*Indicator*	*Mandatory/voluntary disclosure*
36	Environmental	Percentage of GHG emissions reported versus estimated	Mandatory
37	Environmental	Exposure of the benchmark portfolio to companies the activities of which fall under Divisions 05 to 09, 19 and 20 of Annex I to Regulation (EC) No. 1893/2006	Mandatory
38	Environmental	Percentage of green bonds in the benchmark portfolio	Mandatory
39	Social	Weighted average social rating of the benchmark	Voluntary
40	Social	International treaties and conventions, United Nations principles or, where applicable, national law used in order to determine what constitutes a "controversial weapon"	Mandatory
41	Social	Weighted average percentage of benchmark constituents in the controversial weapons sector	Mandatory
42	Social	Weighted average percentage of benchmark constituents in the tobacco sector	Mandatory
43	Social	Number of benchmark constituents subject to social violations (absolute number and relative divided by all benchmark constituents), as referred to in international treaties and conventions, United Nations principles and, where applicable, national law	Mandatory

(continued)

Table 10.7 (continued)

#	*Disclosure area*	*Indicator*	*Mandatory/voluntary disclosure*
44	Social	Exposure of the benchmark portfolio to companies without due diligence policies on issues addressed by the fundamental International Labor Organisation Conventions 1 to 8	Mandatory
45	Social	Weighted average gender pay gap	Mandatory
46	Social	Weighted average ratio of female to male board members	Mandatory
47	Social	Weighted average ratio of accidents, injuries, fatalities	Mandatory
48	Social	Numbers of convictions and amount of fines for violations of anti-corruption and anti-bribery laws	Mandatory
49	Governance	Weighted average governance rating of the benchmark	Voluntary

Table 10.8 ESG disclosures for sovereign debt benchmarks

#	*Disclosure area*	*Indicator*	*Mandatory/voluntary disclosure*
50	Combined ESG factors	Weighted average ESG rating of the benchmark	Voluntary
51	Combined ESG factors	Overall ESG rating of top ten benchmark constituents by weighting in the benchmark	Voluntary
52	Combined ESG factors	The percentage of underlying fund management companies signed up to international standards	Mandatory
52	Environmental	Weighted average environmental rating of the benchmark	Voluntary
53	Environmental	Exposure of the benchmark portfolio to climate-related physical risks, measuring the effects of extreme weather events on companies' operations and production or on the different stages of the supply chain (based on issuer exposure)	Voluntary
54	Environmental	Top ten and bottom ten constituents by exposure to climate-related physical risks	Voluntary
55	Environmental	Greenhouse gas (GHG) intensity of the benchmark	Mandatory
56	Environmental	Percentage of GHG emissions reported versus estimated	Mandatory
57	Environmental	Percentage of green bonds in the benchmark portfolio	Mandatory
58	Social	Weighted average social rating of the benchmark	Voluntary
59	Social	Average human rights performance of the issuers (including a quantitative indicator and the methodology used to calculate it)	Mandatory
60	Social	Average income inequality score, measuring the distribution of income and economic inequality among the participants in a particular economy (including a quantitative indicator and the methodology used to calculate it)	Mandatory

(continued)

Table 10.8 (continued)

#	*Disclosure area*	*Indicator*	*Mandatory/voluntary disclosure*
61	Social	Average freedom of expression score measuring the extent to which political and civil society organizations can operate freely (including a quantitative indicator and the methodology used to calculate it)	Mandatory
62	Governance	Weighted average governance rating of the benchmark	Voluntary
63	Governance	Average corruption score measuring the perceived level of public sector corruption (including a quantitative indicator and the methodology used to calculate it)	Mandatory
64	Governance	Average political stability score, measuring the likelihood that the current regime will be overthrown by the use of force (including a quantitative indicator and the methodology used to calculate it)	Mandatory
65	Governance	Average rule of law score, based on the absence of corruption, respect for fundamental rights and the state of civil and criminal justice (including a quantitative indicator and the methodology used to calculate it)	Voluntary

Table 10.9 ESG disclosures for commodity benchmarks

#	*Disclosure area*	*Indicator*	*Mandatory/voluntary disclosure*
66	Environmental	Degree of exposure of the underlying commodities to climate-related physical risks, measuring the effects of extreme weather events on companies' operation and production or on the different stages of the supply chain (low, moderate or high)	Voluntary
67	Environmental	The methodology used to calculate the climate-related physical risks	Voluntary
68	Environmental	Degree of exposure of the underlying commodities to climate-transition risks, measuring the financial impacts resulting from the effects of the implementation of a low-carbon strategies (low, moderate or high)	Mandatory
69	Social	Degree of exposure of the underlying commodities to social risks (low, moderate or high)	Mandatory
70	Governance	Degree of exposure of the underlying commodities to governance risks (low, moderate or high)	Mandatory
71	Governance	Average rule of law score, based on the absence of corruption, respect for fundamental rights and the state of civil and criminal justice (including a quantitative indicator and the methodology used to calculate it)	Mandatory

Table 10.10 ESG disclosures for benchmarks based on other asset classes

#	*Disclosure area*	*Indicator*	*Mandatory/voluntary disclosure*
72	Combined ESG factors	Weighted average ESG rating of the benchmark	Voluntary
73	Environmental	Weighted average environmental rating of the benchmark	Voluntary
74	Environmental	Degree of exposure of the portfolio to climate-related opportunities, measuring investment opportunities related to climate change, innovating new investment solutions, as percentage of total weight in portfolio	Mandatory
75	Environmental	GHG intensity of the benchmark	Mandatory
76	Environmental	Degree of exposure of the portfolio to the sectors listed in Sections A to H and Section L of Annex I to Regulation (EC) No. 1893/2006 as percentage of total weight in the portfolio	Mandatory
77	Social	Weighted average social rating for the benchmark	Voluntary
78	Social	International treaties and conventions, United Nations principles or, where applicable, national law used in order to determine what constitutes a "controversial weapon"	Mandatory
79	Social	Weighted average percentage of benchmark constituents in the controversial weapons sector	Mandatory
80	Social	Weighted average percentage of benchmark constituents in the tobacco sector	
81	Governance	Weighted average governance rating of the benchmark	Voluntary
82	Governance	Percentage of underlying funds with stewardship policies in place, including measures for the planning and management of resources	Mandatory

avoid the risk of greenwashing, that is the risk that claimed environmental benefits of existing indexes do not actually correspond to real environmental impact. In particular, the Commission was concerned that "*divergent approaches to benchmark methodologies result in the fragmentation of the internal market because it is not clear to users of benchmarks whether a particular low-carbon index is a benchmark aligned to the objectives of the Paris Agreement or merely a benchmark that aims to lower the carbon footprint of a standard investment portfolio*" and that rules should be introduced "*to protect investors from confusion and ambiguity about the aims and level of ambition underpinning different categories of so-called low-carbon indices used as benchmarks for low-carbon investment portfolios*".[14]

The minimum requirements for these two new types of benchmarks were developed by the Technical Expert Group on Sustainable Finance and won't be described here. What is relevant to this discussion is that for investment advisers and retail investors the availability of these official EU index methodologies can be useful in two ways.

Firstly, any passive investment product tracking either an EU Climate Transition Benchmark or an EU Paris-aligned Benchmark will be perceived as immediately compliant with the investment objectives of contributing to the transition to a low-carbon economy. So for example Exchange Traded Funds (ETFs)—a popular retail investment asset class—built around indexes aligned with the official EU Climate methodologies will be reassuring to investment adviser tasked with recommending ESG investment solutions. It's also clear that asset managers manufacturing such products can categorize them as "sustainable investment" (under Article 9 of the SFDR) without fear of greenwashing allegations, as there is a specific mention of this type of investment objective in the level 1 text. It has to be noted that the two EU methodologies apply both to equity benchmarks and to corporate bonds benchmarks.

Secondly, it can be assumed that active investment products falling under benchmarked to either an EU Paris-aligned Benchmark or an EU Climate Transition Benchmark for the purposes of performance attribution will qualify for inclusion in the scope of Article 8 or Article 9 of the SFDR.

The fact that the EU official methodologies have been adopted by asset managers in the EU and beyond even before the entry into application of the regulation suggests that these benefits have positively been perceived by the market.

Product-Level Taxonomy Disclosures

The role of the Taxonomy in the EU Action Plan has already been discussed in Chapter 4. What has to be noted here is that Taxonomy disclosures complement the disclosures under Article 8 and 9 of the SFDR. The product level disclosures outlined in the level 1 text are summarized in Table 10.11.

In March 2021, the ESAs launched a consultation[15] with the following objectives:

1. To develop additional disclosure rules on the pre-contractual information for Article 8 products under the SFDR using the environmental taxonomy: these are so-called "light green" products—i.e. those that promote environmental characteristics, leading to technical standards for climate objectives and other environmental objectives.
2. To develop additional disclosure rules on the pre-contractual information for Article 9 products under the SFDR: these are the so-called "dark green" products—i.e. those that invest in an "economic activity that contributes to an environmental objective". The criteria will make use of the environmental taxonomy, leading to

Table 10.11 EU Taxonomy disclosures at the product level (Articles 4 and 5)

#	*Disclosure area*	*Indicator*	*Metric*
1	Transparency of environmentally sustainable investments in pre-contractual disclosures and in periodic reports	The information on the environmental objective or environmental objectives to which the investment underlying the financial product contributes	–
2	Transparency of environmentally sustainable investments in pre-contractual disclosures and in periodic reports	A description of how and to what extent the investments underlying the financial product are in economic activities that qualify as environmentally sustainable	1. The proportion of investments in environmentally sustainable economic activities selected for the financial product 2. Details on the proportions of enabling and transitional activities as a percentage of all investments selected for the financial product

technical standards for climate objectives and other environmental objectives.

3. To develop additional disclosure rules on the periodic information for both Article 8 and Article 9 products under the SFDR making use of the environmental taxonomy, leading to technical standards for climate objectives and other environmental objectives.

While the regulatory process is still underway at the time of writing,[16] it has to be noted here that Taxonomy alignment disclosures, similar to other disclosure discussed above, will be of limited usefulness from an investment decision perspective. There is currently in the level 1 text no minimum requirement for alignment with Taxonomy even for "dark green" products under Article 9 of the SFDR. Any percentage (see indicator 2 and associated metrics in Table 10.6) in terms of portfolio assets aligned with the Taxonomy is acceptable, however small. Therefore, the indicator can be relevant only in a comparison. It can be assumed that the greater the percentage of assets aligned with Taxonomy, the greener the fund or investment product. However, the methodology used to construct and manage the portfolio could encompass further criteria—e.g. engagement or stewardship strategy with regard to carbon intensive asset, best-in-class selection of assets belonging to high-carbon sector, etc.—which might make a fund with lower exposure to Taxonomy sectors more impactful and ambitious than a fund with a higher Taxonomy alignment.

The Action Plan ultimately leaves the judgement of a fund's sustainability effectiveness to investors—and prudently so it could be argued. In parallel, for those investors who require more guidance on what assets can qualify as green, another tool—directly built on Article 5 and 6 Taxonomy disclosure—has been developed: the EU Ecolabel for financial products.

The EU Ecolabel for Financial Products

One of the most complex measures of the EU Action Plan,[17] the EU Ecolabel for financial products is expected to be launched in the last quarter of 2021. Led by the Joint Research Centre of the European Commission, the work for the development of the label criteria has seen a number of proposals put forward for consultation over the course of 2019 and 2020. The criteria discussed here refer to the version released in October 2020.[18]

The Ecolabel criteria cover fundamentally two product categories:

1. The service of managing an investment product that has been packaged for retail investors in accordance with the requirements laid down in Regulation (EU) No. 1286/2014 on packaged retail and insurance-based investment products (PRIIPs). This shall include:
    - Equity, bond and mixed investment funds, to include those referred to as Undertakings for the Collective Investment in Transferable Securities (UCITS) and Alternative Investment Funds (AIFs) that are marketed to retail investors upon a national discretion (according to Article 43 of Directive 2011/61/EU);
    - Insurance-based investment products (IBIPs), to include unit-linked, profit participation and multi-option life insurance products.

   Units or shares in European Long Term Infrastructure Investment Funds (ELTIFs) and real estate funds can be present in the underlying assets of a Retail AIF or insurance product applying for an EU Ecolabel.
2. The service of managing a fixed-term deposit or savings deposit product as referred to in Article 2(1) point 3 of Directive 2014/49/EU on deposit guarantee schemes. The service shall be provided by the credit institution on whose balance sheet the deposits held (liabilities) and the associated loans granted as credits (assets) appear.

Criteria are then developed for each product type, and in particular for

(a) Retail AIF funds
(b) UCITS equity funds
(c) UCITS bond funds
(d) UCITS mixed funds
(e) Funds of Funds
(f) Unit-linked insurance products
(g) Profit participation insurance products
(h) Multi-option and hybrid insurance products
(i) Other assets

For all the products, the criteria are:

1. Investment in environmentally sustainable economic activities
2. Investment in companies investing in transition and green growth
3. Exclusions based on environmental aspects
4. Exclusions based on social and governance aspects
5. Engagement
6. Measures taken to enhance investor impact
7. Retail investor information
8. Information appearing on the EU Ecolabel

Just to take the UCITS equity fund example, the first criterion is laid down as follows: at least 40% of the total portfolio value of the assets under management (AuM) shall be invested in environmentally sustainable economic activities, which may include a contribution from companies investing in transition or green growth, according to the formula below[19]:

$$G = \Sigma(\mathrm{PCi} * (0.6 * \mathrm{GRi} + 0.4 * \mathrm{GCi})), \quad \text{if } \mathrm{GCi} > \mathrm{GGRini} = 1$$
$$G = \Sigma(\mathrm{PCi} * (0.6 * \mathrm{GRi} + 0.4 * \mathrm{GRGi})), \quad \text{if } \mathrm{GCi} < \mathrm{GGRini} = 1$$

where

G	% of total portfolio value invested in environmentally sustainable economic activity
i	an individual company in which portfolio equities are held
n	total number of companies in the portfolio
PCi	% Portfolio contribution of company *i*
GRi	% Green Revenue of company *i*
GCi	% Green Capex
GRGi	cumulative % projected Green Revenue Growth *i* over 5 years.

The formula means that "each company's percentage contribution shall be calculated as the weighted average of their % green revenue (GRi) and their % Green Capex (GCi) or % projected Green Revenue Growth (GRGi). The company contribution to the total portfolio value is then weighted based on the % of holdings in the total portfolio value (PCi). The sum of the weighted company contributions gives the % of total portfolio value invested in environmentally sustainable economic activity". The reference to the Taxonomy sits in the definition of the activities qualifying as sustainable.[20]

Without exploring the Ecolabel draft criteria in further detail, it's clear from the list of products in scope that it will only cover funds and not financial instruments under MiFID II and the IDD. If adopted by the market, it's likely to be of use to investment advisers and retail investors in signalling dark green products, certified through hard criteria and a strong verification process. The certification will then be complemented by the disclosures under SFDR Article 9, including Taxonomy related, both in pre-contractual documents and in periodic reports.

The EU Green Bond Standard

The only regulatory tool that addresses the issue of signalling the ESG relevance of single transferable securities is the EU Green Bond Standard. An initiative of the EU Action Plan whose details were developed by Technical Expert Group on Sustainable Finance,[21] a proposal for a Standard for European Green Bonds was released for consultation on 6th July 2021.[22] The Commission had already consulted the market and a wide set of stakeholders on its establishment through an ad-hoc consultation[23] and as part of the broader consultation on the renewed sustainable finance strategy[24] in 2020.

The EU GBS is designed as a voluntary standard, that can be adopted by issuers of all types (including both listed and privately held), with the objective of strengthening the criteria for the qualification of bonds as sustainable compared to existing international frameworks, such as those issued by the International Capital Markets Association (ICMA).

Its awarding criteria cover four areas:

(i) The alignment of the use-of-proceeds with the EU Taxonomy;
(ii) The content of a Green Bond Framework to be produced by the issuer;
(iii) The required Allocation and Impact Reporting; and
(iv) The requirements for external verification by an approved verifier.

While the introduction of an EU framework can help to create consistency across different regulatory interventions (especially by establishing a further use-case for the EU Taxonomy), its additionality from the perspective of an investment adviser or a retail investor is questionable. The market process for the recognition of green bonds is already well established and has rarely given rise to market failures or reputational controversies, since it already relies on the crucial role of independent providers of verification services. It is also limited to use-of-proceeds bonds excluding other emerging categories of sustainability bonds (such as sustainability-linked bonds or transition bonds, which rely on the articulation of a long-term transition strategy on the part of the issuer). However the Renewed Sustainable Finance Strategy, published by the Commission on 6th July,[25] envisages the introduction of new bond *labels* by 2022 (such as the *transition label* or the *sustainability-linked bond label*).

Case Study: The Draft Templates for Product Disclosures Under the SFDR

In September 2020, the ESAs launched a survey on environmental and/or social financial product templates.[26] In particular, illustrative mock-ups of disclosures for products falling within the scope of Article 8 of the SFDR were published, covering a fictional case of a passive investment fund tracking an ESG index, with a view to standardizing "the disclosures by requiring the use of specific templates, recognising the need for standardisation of disclosures to promote comparability of different financial products in different Member States with respect to ESG information, in line with Recital 9 of the SFDR".[27]

The proposed fictional example is an interesting case-study, as it shows what all the proposed disclosures will look like or are expected by the regulator to look like, beyond the dry and sometime awkward legal wording of the Regulatory Technical Standards.

Here is presented and commented the content of "Illustrative Mock up 1: Pre contractual disclosures of a product with E/S characteristics (with icons)"[28] (Table 10.12).

This section implies that the fundamental distinction—between having ESG characteristics and having an ESG objectives—has become straightforward and familiar to both investment advisers and retail investors, without requiring further explanations (Table 10.13).

The ESG strategy of the financial product is passive and combines a best-in-class approach with sector exclusions (Table 10.14).

Table 10.12 Excerpt from ESAs illustrative mock-up 1: Product labelling

Environmental and/or social characteristics
This product has ☐ A sustainable objective ☒ ESG characteristics This financial product promotes environmental or social characteristics, but does not have as its objective a sustainable investment. It might invest partially in assets that have a sustainable objective, for instance qualified as sustainable according to the EU classification

Table 10.13 Excerpt from ESAs illustrative mock-up 1: example description of ESG characteristics

What environmental and/or social characteristics are promoted by this financial product?
The characteristics promoted by this financial product consist of investing in corporates with excellent Environmental, Social and Governance ratings while excluding issuers producing nuclear power, owning fossil fuel reserves, producing tobacco or military weapons as well as any company from the gambling sector. This product is passively managed. Its objective is to track the performance of the TRXI World 5% Issuer Capped ESG Index, and to limit to max. 1% the tracking error between the net asset value of the product and the performance of the Index

The most important additional information captured here compared to the previous section regards the assessment of governance practices of investee companies. As per the level 1 text, an Article 8 product is one that promotes among other characteristics, environmental or social characteristics or a combination of those characteristics, "provided that the companies in which the investments are made follow good governance practices". What is striking in the example provided here is that the fact that the companies are rated for governance aspects is sufficient disclosure, and no expectation is set for the companies to meet a minimum governance score (theoretically, even companies with low governance ratings could be included, as far as the disclosures provided here are concerned) (Table 10.15).

What is striking here is that it is sufficient for the asset manager to state that 15% of the fund holdings are "qualified as sustainable", without need for providing further explanations. It is also enough to say that

Table 10.14 Excerpt from ESAs illustrative mock-up 1: investment strategy

	What investment strategy does this financial product follow?
An investment strategy is what guides the decision on what to invest in based on e.g. investment goals, risk tolerance and future needs for liquidity **E/S** stands for "environmental and/or social" **Binding elements** are commitments that cannot be amended during the life of the financial product **Governance** is about a company's leadership, remuneration of staff (incl. executives), audits, internal controls, shareholder rights, tax	The investment strategy aims at tracking the TRXI World 5% Issuer Capped ESG Index. The TRXI World 5% Issuer Capped ESG Index applies a max 5% weight to the largest holdings of the TRXI World ESG Index. The latter captures companies with excellent Environmental, Social and Governance ratings, following a best-in-class investment strategy, while excluding issuers producing nuclear power, owning fossil fuel reserves, producing or selling tobacco or military weapons as well as any company from the gambling sector. Both the TRXI World 5% Issuer Capped ESG Index and the TRXI World ESG Index serve as a potential replacement for current TRXI World index exposure for investors interested in selecting corporates with the best ESG ratings while eliminating companies that have a negative impact on certain sustainability factors from their portfolio **What are the binding elements for the investment selection?** This product is aligned with the TRXI World 5% Issuer Capped ESG Index at 95%. The TRXI World 5% Issuer Capped ESG Index, tracked by this product, targets securities of companies with the highest ESG ratings making up 25% of the market capitalization in each sector and region of the parent index. Companies must have an RTT ESG rating above "BBB" and the RTT ESG Controversies score greater than 4 to be eligible for the TRXI ESG Index. The Index applies a maximum 5% weight to the largest holdings. The exclusion of issuers producing nuclear power, owning fossil fuel reserves, producing tobacco or military weapons as well as any company from the gambling sector is another binding element

(continued)

Table 10.14 (continued)

	What investment strategy does this financial product follow?
	How is the strategy implemented in the investment process on a continuous basis? The investment seeks to provide investment results that, before fees and expenses, correspond generally to the total return performance of the TRXI World 5% Issuer Capped ESG Index. The fund generally invests substantially all, but at least 95%, of its total assets in the securities comprising the index. In addition, it may invest in equity securities that are not included in the index, cash and cash equivalents or money market instruments. The fund is non-diversified. Compliance with the characteristics is monitored on a regular basis **Is there an amount of potential investments excluded, as a result of the implementation of the selection criteria?** ☐ No ☒ Yes, the implementation of the selection criteria leads to the exclusion of at least 50% of potential investments **What is the policy to assess good governance practices of the investee companies?** The investee companies are rated for governance aspects using the RTT ESG Ratings (www.RTT-ESG-Rating.com)

Table 10.15 Excerpt from ESAs illustrative mock-up 1: minimum asset allocation

	What is the minimum asset allocation planned for this product?
The category #1 Aligned with E/S characteristics includes any investment that is aligned with the characteristic:- **The Subcategory #1A** covers investments that are qualified as sustainable **The sub-category #1B Other E/S characteristics** covers investments aligned with the environmental or social characteristics that do not meet the criteria of "sustainable investment".- **#2Remainder** includes investments which are not aligned with the E/S characteristics, nor are qualified as sustainable	The fund invests in direct holdings. In order to meet the environmental or social characteristics promoted, the fund generally invests at least 95% of its total assets in the securities comprising the TRXI World 5% Issuer Capped ESG Index. Hence, 95% of the investments are aligned with the environmental characteristics. This includes 15% of the total investments that are qualified as sustainable, some of which are classified as environmental investments under the EU framework to facilitate sustainable investment). 5% of the total investments do not incorporate any environmental or social characteristic

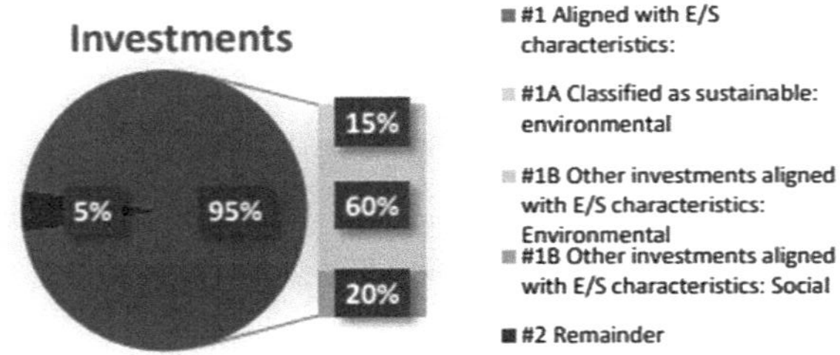

(continued)

Table 10.15 (continued)

What is the minimum asset allocation planned for this product?

In which economic sectors are the investments made?
The fund is mostly exposed to the Information Technology and Health Care sectors

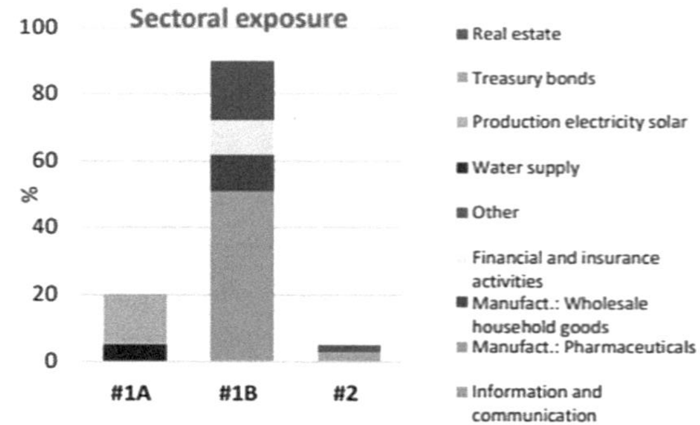

"some" of those sustainable investments are "classified as environmental investments" under the EU Taxonomy (Table 10.16).

To note here that no reference is made to the % of the assets that is compliant with the EU Taxonomy, it is only mentioned that investment in solar photovoltaic energy production is Taxonomy aligned, but without any reference to how the Technical Screening Criteria for substantial contribution are met. There is instead a reference to Do No Significant Harm criteria for photovoltaic solar energy production and water supply.

While the template was created to outline a possible structure of product disclosures under the SFDR rather than to set a standard for its content, the fact that the ESAs present its content in this way seems to suggest that flexibility around the application of the SFDR is a priority in the context of an EU Action Plan and an EU disclosure regulatory environment still under development. In the absence of clear guidance regarding the content of the EU Taxonomy, and in the face of data availability challenges, the message sent across by the ESAs seems to be that narrative disclosures based on information that can be easily sourced should suffice to inform end investors regarding the sustainability characteristics of an investment product.

Concluding Remarks

The toolbox developed for the retail investment use-case as part of the EU Action Plan has introduced significant innovations in the EU sustainable finance regulatory environment. However, it presents certain gaps and inconsistencies that can be addressed either by market-based solutions, or by future regulatory interventions (for example, as part of the next iteration of the EU sustainable finance strategy in the context of the EU Green Deal).

The most significant gap is represented by the different scope of application of the delegated regulation on sustainability consideration in the suitability test under MiFID II on the one hand, and the definition of sustainable investment and products with ESG characteristics under the SFDR. The analysis carried out in this Chapter highlights how financial instruments such as equity shares, corporate bonds (other than green bonds) and sovereign bonds are not covered by any regulatory guidance with regard to their eligibility for investment advice where preferences have been expressed for sustainable products or products with ESG characteristics.

Table 10.16 Excerpt from ESAs illustrative mock-up 1: substantial contribution and do-no-harm information

	How will sustainable investments contribute to a sustainable objective and not significantly harm any other sustainable investment objectives during the reference period?
This section only concerns the portion of sustainable **investments illustrated above (#1A Sustainable),** and not the other E/S characteristics part, nor the remainder An investment that contributes to an environmental or social objective can only be considered "sustainable" where it demonstrates a contribution to an environmental or social objective, and it **does not significantly harm** any (other) sustainability objective	**How will sustainable investments of the financial product contribute to a sustainable objective?** Investments in solar photovoltaic energy production are considered as significantly contributing to climate change mitigation under the EU Taxonomy. The investments in companies supplying water contribute to climate change adaptation. The activity of the company supplying water has a substantial contribution to GHG emissions savings through low specific energy consumption in the water supply system measured in kWh per cubic meter of water. Nonetheless, all of the selected companies supplying water might not be compliant with EU Taxonomy energy consumption thresholds for substantial contribution to climate change adaptation. Our applied threshold is that of 1 kWh per cubic meter of water, instead of 0.5 kWh as per the EU Taxonomy

	How will sustainable investments contribute to a sustainable objective and not significantly harm any other sustainable investment objectives during the reference period?
	How are principal adverse impacts on sustainability factors taken into account? Upon investment and over the life of the product, we assess and monitor indicators that are deemed to indicate the presence of a principal adverse impact as per EU law, except for all biodiversity-related indicators, for which we are unable to collect data. More details can be found under the prospectus section on Adverse Impact. We address adverse impacts by engaging with investee companies. We use research from proxy voting companies to help us decide how to vote **How are investments excluded that significantly harm sustainable investment objectives?** All investments qualifying as sustainable (15%) are screened against all significant harm indicators relevant to solar photovoltaic energy production and water supply under the EU Taxonomy, with the exception of biodiversity-related indicators. Environmental Impact Assessment have been completed were relevant. The implementation of site-level biodiversity management plan complies with the IFC Performance Standard 6: Biodiversity Conservation and Sustainable Management of Living Natural Resources. The production of electricity from solar Photovoltaic did not substantially undermine climate change mitigation

It can be expected that the market will resort to ESG ratings—where available—as a market-based solution to address the emerging ESG categorization needs for equities. It has to be noted however that ESMA has recently expressed concerns over such unregulated tools,[29] highlighting "the need to match the growth in demand for these products with appropriate regulatory requirements to ensure their quality and reliability".

On the other hand, the integration of ESG consideration into credit rating—already regulated by ESMA—can meet the ESG categorization needs for corporate and sovereign bonds that do not fall in the green or sustainable bond category.

Regarding this asset class, it has been remarked that the introduction of an EU Green Bond Standard does not seem to be a priority, given the availability of reliable and well-established market-based standards.

The main other area of concern is the ability of product disclosures under Articles 6, 8 and 9 of the SFDR to provide investment advisers and retail investors with meaningful and decision-useful information that can orient investment decisions in a direction consistent with investment preferences.

In this respect, the reliance of these disclosures on the presence of an ESG benchmark highlight the need to focus the attention of policy makers in the next iteration of the EU Sustainable Finance strategy in a review of both the ESG disclosure for benchmarks and the EU Climate benchmarks methodologies to ensure that they are fit for purpose in the light of the dependencies arising from other measures resulting from the EU Action Plan.

Notes

1. See European Commission (2018, p. 4): "Clear guidance on activities qualifying as contributing to climate change mitigation and adaptation, environmental and social objectives will help inform investors. It will provide detailed information on the relevant sectors and activities, based on screening criteria, thresholds and metrics. This is an essential step in supporting the flow of capital into sustainable sectors in need of financing".
2. As discussed in Chapter 5, the 2017 revision of the Shareholder Rights Directive (Directive (EU) 2017/828) aims to encourage long-term shareholder engagement "to ensure that decisions are made for the long-term

stability of a company and take into account environmental and social issue".

3. For a detailed analysis of all these measures, a useful reference is EFRAG European Lab (2021). See also Chapter 4 of this book.
4. In 2020 the European Commission launched a consultation on its renewed sustainable finance strategy. The new strategy is expected to be adopted by June 2021.
5. European Commission (2018, p. 7).
6. Commission Delegated Regulation (EU) 2021/1253 of 21 April 2021 amending Delegated Regulation (EU) 2017/565 as regards the integration of sustainability factors, risks and preferences into certain organisational requirements and operating conditions for investment firms (Text with EEA relevance.
7. Article 1 (12) of Regulation (EU) 2019/2088.
8. See ESAs (2020).
9. See ESAs (2021a, p. 13).
10. See Recital 18.
11. Paris-alignment disclosures instead are also applicable to standard benchmarks without an ESG objective.
12. See "Delegated acts supplementing Regulation (EU) 2016/1011 of the European Parliament and of the Council as regards the explanation in the benchmark statement of how environmental, social and governance factors are reflected in each benchmark provided and published".
13. Regulation (EU) 2019/2089.
14. Ibidem, Recital 12.
15. See ESAs (2021b).
16. Following the completion of the consultation phase, the ESAs were expected to publish their final report by July 2021.
17. The action regarding the introduction of a so-called Green Supporting Factor might qualify as even more challenging.
18. Joint Research Centre (2020).
19. Annex to Joint Research Centre (2020, p. 6).
20. The full list of environmentally sustainable activities is expected to be adopted by the Commission by the end of 2021.
21. For the final report on the EU Green Bond Standard, see Technical Expert Group on Sustainable Finance (2020).
22. Text available at https://eur-lex.europa.eu/legal-content/EN/TXT/?uri=CELEX:52021PC0391.
23. "Targeted consultation on an EU Green Bond Standard", accessible at https://ec.europa.eu/info/consultations/finance-2020-eu-green-bond-standard_en.
24. "Consultation on the renewed sustainable finance strategy", accessible at https://ec.europa.eu/info/consultations/finance-2020-sustainable-finance-strategy_en.

25. For more details see https://ec.europa.eu/commission/presscorner/detail/en/ip_21_3405.
26. The press release is retrieved from https://www.esma.europa.eu/press-news/esma-news/esas-launch-survey-environmental-andor-social-financial-product-templates.
27. See introduction to the survey at https://ec.europa.eu/eusurvey/runner/ESGtemplatesSFDR.
28. The template can be downloaded at the link above.
29. For the full text of the letter see ESMA (2021).

References

CONSOB. (2020). *Report on financial investments of Italian households Behavioural attitudes and approaches*. ccfe7ad2-810f-4490-bd7e-413daa24c391 (consob.it)

EFRAG European Reporting Lab. (2021). *EU non-financial reporting momentum and coherence*. https://www.efrag.org/Assets/Download?assetUrl=%2Fsites%2Fwebpublishing%2FSiteAssets%2FEFRAG%2520PTF-NFRS_A1_FINAL.pdf

European Commission. (2018). *Action plan: Funding sustainable growth*. https://eur-lex.europa.eu/legal-content/EN/TXT/?uri=CELEX:52018DC0097

ESAs. (2020). *Joint ESA consultation on ESG Discosures*. https://www.esma.europa.eu/press-news/esma-news/esas-consult-environmental-social-and-governance-disclosure-rules

ESAs. (2021a). *Final report on draft regulatory technical standards with regard to the content, methodologies and presentation of disclosures pursuant to Article 2a(3), Article 4(6) And (7), Article 8(3), Article 9(5), Article 10(2) And Article 11(4) of regulation (EU) 2019/2088*. https://www.esma.europa.eu/sites/default/files/library/jc_2021_03_joint_esas_final_report_on_rts_under_sfdr.pdf

ESAs. (2021b). *Joint consultation paper: Taxonomy-related sustainability disclosures*. https://www.esma.europa.eu/press-news/consultations/joint-consultation-taxonomy-related-sustainability-disclosures

ESMA. (2021). *Letter to the European Commission on ESG ratings*. https://www.esma.europa.eu/press-news/esma-news/esma-calls-legislative-action-esg-ratings-and-assessment-tools

Joint Research Centre. (2020). *Development of EU ecolabel criteria for retail financial products*. https://susproc.jrc.ec.europa.eu/product-bureau/sites/default/files/2020-10/Draft%20Technical%20Report%203%20-%20Retail%20financial%20products.pdf

Technical Expert Group on Sustainable Finance. (2020). *EU green bond standard usability guide*. https://ec.europa.eu/info/sites/info/files/business_economy_euro/banking_and_finance/documents/200309-sustainable-finance-teg-green-bond-standard-usability-guide_en.pdf

CHAPTER 11

Conclusions

Claudia Guagliano, Nadia Linciano, and Paola Soccorso

The health emergency has reawakened and strengthened the attention of policymakers to sustainable development, reaffirming the urgency of economic policies aimed at mainstreaming economic, environmental and social dimensions to promote a resilient system (McDaniels, 2020). That is particularly felt in the context, where ecological transition is both a tool and an objective of the actions put in place to foster economic recovery, such as the NGEU programme.

C. Guagliano
The European Securities and Markets Authority, Paris, France
e-mail: claudia.guagliano@esma.europa.eu

N. Linciano (✉)
Commissione Nazionale per le Società e la Borsa, Rome, Italy
e-mail: n.linciano@consob.it

P. Soccorso
Commissione Nazionale per le Società e la Borsa, Rome, Italy
e-mail: p.soccorso@consob.it

N. Linciano et al. (eds.), *Information as a Driver of Sustainable Finance*, Palgrave Studies in Impact Finance,
https://doi.org/10.1007/978-3-030-93768-3_11

At the same time, the pandemic reinforced the already increasing investors' attitude towards consideration of ESG factors. In the last quarter of 2020, as already mentioned, net inflows into Eurozone-domiciled equity ESG funds reached an all-time high (ESMA, 2021). In addition, during the height of financial market turbulence, ESG funds and green bonds performed well or better than comparable assets (ESMA, 2021).

The crisis triggered by COVID-19 has also brought to the fore the effects that catastrophic and global events can have on the sectors most exposed to ESG risks. Sharp corrections in market valuations of fossil fuel companies and non-renewable energy producers, led also by the acceleration of decarbonisation driven by stricter regulation, may hinder financial stability. This has led to a renewed need for stress tests and contingency plans in the event that a shock to market valuations generates systemic impacts.

At the other extreme, challenges are posed by the risk of overvaluation of financial instruments qualifying as ESG investments, due to investors' growing interest in sustainability.

The risk of green and social washing is also increasingly perceived, as evidence on the proportion of the phenomenon starts to be gathered. In January 2021, greenwashing in a non-financial context was for the first time the focus of the annual screening of websites ("sweep") conducted by the European Commission to identify breaches of EU consumer law in online markets. The 'sweep' analysed green online claims from various business sectors such as garments, cosmetics and household equipment. The investigation found that in more than half of the cases, the company had not provided consumers with sufficient information and data to assess the truthfulness of claims about the 'green' characteristics of its business on its website. The situation is not too dissimilar in the financial sector: according to a recent analysis by InfluenceMap (2021), out of 593 equity funds in the broad ESG category with over $265 billion in total net assets '*421 (equivalent to 71% of the sample), have a negative Portfolio Paris Alignment score, indicating the companies within their portfolios are misaligned from global climate targets*'. There is still a lot to be done on this issue.

By now, it is therefore clear that sustainability and ESG factors pose both opportunities and risks, to be governed also through the development of a well-functioning information ecosystem. All the players involved in the production and dissemination of information as well as the market infrastructures allowing the incorporation of non-financial information into market prices have to be involved.

The digitalisation of finance (so-called FinTech[1]) can play an important role in the development of sustainable finance, in a combination often referred to as Green FinTech.

The topic has been under the attention of some international institutions for some time. Since 2016, several United Nations reports have highlighted the key role that technology can play in granting: the availability of more accurate, timely and reliable information on the ESG characteristics and ESG impact on businesses and financial products; the accessibility of ESG information also to retail investors; wider opportunities and channels of access to capital markets to the benefit of both investors and companies, thanks to the disintermediation resulting from the development of digital platforms dedicated to sustainability (UNEP, 2016; UN Secretary General's Task Force on Digital Financing of the Sustainable Development Goals, 2019a, 2019b).[2]

OECD (2019) mentions the application of distributed ledger technology (DLT), artificial intelligence (AI), machine learning and the internet of things (IoT) to the assessment of ESG performance and to the exposure to ESG risks of a specific business project, a company, one or more undertakings in a supply chain, a portfolio of equities or bank loans.[3]

Green FinTech would benefit transparency, fairness and comparability of information as well as the efficiency of financial markets. This would mitigate the issue particularly felt by the community of institutional investors, asset managers and banks, who often point to the lack of reliable, comparable and usable data for scenario analysis as the most significant obstacle to integrating sustainability into their decision-making processes.[4]

The application of DLT technology to green bond issuance is another area of growing interest, although the scale of the phenomenon is still very small. The first blockchain-based bond issuances exhibit significant cost savings in the design, issuance and distribution, transfer of ownership and settlement, and reporting on the use of the funds. In perspective, therefore, blockchain-based green (or sustainability-related) bonds could be attractive to both SMEs (with reduced costs compared to normal bond issues) and investors engaged in sustainability (as information on the environmental impact of the project financed by the bond is accessible). The digitalisation of bond issues also paves the way for product innovations that could boost participation in capital markets and, in particular, in the sustainable investment sector (HSBC, 2019).

In addition, in the area of investment services and asset management, many see robo-advisory services and the development of automated tools to detect and integrate ESG preferences as a way to steer savers towards sustainable investments (ESMA, 2021; Macchiavello & Siri, 2020).

Finally, the use of technology in the compliance function by supervised entities (so-called RegTech) and in supervisory activities by regulators (so-called SupTech) could facilitate, respectively, the implementation and supervision of sustainability-linked obligations, such as those following the 2018 Action Plan in the European context (Macchiavello & Siri, 2020).

The potential of Green FinTech goes hand in hand with some concerns. The first refers to the consumption of energy and resources that characterise blockchain, big data analysis etc. A possible solution could be the use of DLT mechanisms that foresee a lower energy consumption, as they replace the proof-of-work validation method with the proof-of-authority validation method, based on a lower energy-consumption architecture (OECD, 2019).

In the field of investor protection, some critical issues already debated with regard to the digitalisation of investment services are also relevant, e.g. in terms of algorithm opacity, user discrimination, privacy protection and cybersecurity. If the widespread diffusion of FinTech were to result in the exclusion of segments of the population with lower financial and digital skills, a negative social effect would emerge that would conflict with the sustainability goals that can be achieved through digitalisation (Macchiavello & Siri, 2020).[5]

Finally, the development of Green Fintech (and FinTech in general) may be hindered by the absence of a clear and homogeneous regulatory framework across countries, capable of resolving the uncertainties that financial innovation generates on multiple levels. At the European level, the European Commission's Digital Finance Package includes legislative proposals concerning, among other things, a pilot regime for DLT-based market infrastructures (Pilot Regime Regulation) with the aim of establishing a harmonised regulatory framework. At the global level, collaboration and sharing between jurisdictions is being fostered by initiatives such as the Global Blockchain Policy Forum, the Blockchain Policy Centre and the Sustainable Infrastructure Policy Initiative of the OECD

(2019) and the Network for Greening the Financial System (NGFS) promoted by central banks.

* * * * *

The European legislator has intervened with a wide-ranging project, which takes the lead amid market developments impaired by obstacles largely linked to a lack of reliable and comparable data and information. The legislative measures adopted and in the process of being adopted at the EU level act on several fronts, first of all by defining a common language, extending the number of market players involved (for example, by assigning an explicit role to the distributors of financial products, which until now have been less involved than institutional investors), reviewing existing rules in the light of application experience (as in the case of non-financial reporting), and intervening in terms of transparency and rules of conduct.

As with all financial market regulations, however, and even more so in light of the global nature of environmental issues, global coordination is needed to ensure the development of principles that are as shared as possible and not specific to individual areas. It is no coincidence that the Strategy for financing the transition to a sustainable economy, adopted last July, emphasises the need for an ambitious and robust international architecture for sustainable finance, in the understanding that global challenges require global action.

The effectiveness of sustainable finance measures in the EU also depends on their consistent application across the Member States. The European supervisory authorities, in cooperation with national supervisory authorities, play a key role in ensuring convergence of rules and supervisory practices, also by issuing standards and conducting peer reviews.

In the area of sustainability, as in other areas, there is also a risk that overly broad and complex regulation may lead to excessive burdens on market participants (including companies and intermediaries). This risk is particularly relevant for SMEs, for which it is essential to apply a proportionality principle.

Finally, it is important that new regulations are perceived not only as a cost but also as an opportunity to trigger virtuous behaviour and stimulate a cultural transformation both in supervised entities and in supervisors themselves.

This is a necessarily gradual, challenging path for all the actors involved, including the Financial Market Supervisory Authorities. These have the task of accompanying market participants towards mainstreaming ESG factors into their choices according to an approach of substantive compliance, which is an indispensable precondition for the transformation of business models towards a long-term vision. To this end, raising awareness among all market participants (e.g. SMEs with regard to non-financial reporting) and education of small investors is key. The latter is a theme explicitly referred to in the new CMU Action Plan[6] where it refers to the objective of '*Making the European Union an even safer place for people to save and invest for the long term*', providing in action 7 a feasibility analysis for the development of a European framework of financial skills and the possibility for states to promote '*learning measures in support of financial education, in particular in relation to responsible and long-term investment*'.

In order to best address the transition to the sustainable finance paradigm, it is increasingly important that supervisory authorities adopt a proactive and evidence-based approach, as part of an ongoing dialogue with their stakeholders (primarily savers and the financial community at large).

The mainstreaming of sustainability in the decision-making processes of economic and financial actors is a unique opportunity to strengthen the servant role of finance vis-à-vis the real economy, which cannot be missed.

Notes

1. According to the 2018 EC FinTech Action plan, '*FinTech is a term used to describe technology-enabled innovation in financial services that could result in new business models, applications, processes or products and could have an associated material effect on financial markets and institutions and how financial services are provided*' (see the Communication from the Commission to the European Parliament, the Council, the European Central Bank, the European Economic and Social Committee and the Committee of the Regions, FinTech Action plan: For a more competitive and innovative European financial sector, Brussels, 8.3.2018 COM(2018) 109 final; p. 2. Retrieved from https://eur-lex.europa.eu/resource.html?uri=cellar:6793c578-22e6-11e8-ac73-01aa75ed71a1.0001.02/DOC_1&format=PDF).
2. We refer, in particular, to crowdfunding, which takes shape through the use of highly specialised platforms and constitutes a concrete way of financial inclusion of small businesses engaged in activities in line with the SDGs,

whose credit needs are not met by traditional credit channels. See Macchiavello and Siri (2020) also for references to initiatives already launched in several countries, including Italy.

3. BNP Paribas (2019a) illustrates how advances in computer computing power and the application of maching learning to the collection and analysis of geospatial information make data on the environmental impacts (such as CO2 emissions and deforestation) of an economic activity and an individual asset accessible in an accurate and timely manner.
4. See, among others, the 2019 Survey conducted by BNP Paribas (2019b) on a sample of 347 institutional investors. Respondents are engaged in the development of systems, including proprietary artificial intelligence and big data analytics, ESG reporting and benchmarking, to more accurately assess the ESG performance of the companies in which they have invested and to compare it in real time with that of similar companies, their sector or their benchmark.
5. This conflict is part of the broader theme of trade-offs that may exist between the different SDGs: for example, the transition to a green economy and the consequent decarbonisation results in the strong containment of fossil fuel extraction and use with income and wealth distributional effects that may hinder the overcoming of poverty (Campagnolo & Davide, 2019).
6. See the Communication from the Commission to the European Parliament, the Council, the European Central Bank, the European Economic and Social Committee and the Committee of the Regions https://eur-lex.europa.eu/resource.html?uri=cellar:61042990-fe46-11ea-b44f-01aa75ed71a1.0010.02/DOC_1&format=PDF. Retrieved from https://eur-lex.europa.eu/resource.html?uri=cellar:6793c578-22e6-11e8-ac73-01aa75ed71a1.0001.02/DOC_1&format=PDF.

References

BNP Paribas. (2019a). *FinTechs and the ESG data challenge—Six case studies of emerging technologies*. https://securities.bnpparibas.com/files/live/sites/web/files/medias/documents/thoughtleadership/art_fintechs_esg_2019a-09-09.pdf

BNP Paribas. (2019b). *ESG gobal survey 2019*. https://securities.bnpparibas.com/files/live/sites/web/files/medias/documents/esg/esg-global-survey-en-2019.pdf

Campagnolo, L., & Davide, M. (2019). Can the Paris deal boost SDGs achievement? An assessment of climate mitigation co-benefits or side-effects on poverty and inequality. *World Development, 122*, 96–109. https://www.sciencedirect.com/science/article/abs/pii/S0305750X19301299

ESMA. (2021). *Report on trends, risks and vulnerabilities, no. 1*. https://www.esma.europa.eu/document/esma-report-trends-risks-and-vulnerabilities-no1-2021

European Commission. (2021,January 28). *Screening of websites for 'greenwashing': Half of green claims lack evidence*. Press release. https://ec.europa.eu/commission/presscorner/detail/en/ip_21_269

HSBC. (2019). *Blockchain gateway for sustainability linked bonds*. https://docs.wixstatic.com/ugd/3d4f2c_a8d74cac6f974984a67f7b8c05146fbe.pdf

InfluenceMap. (2021). *Asset manager and climate change: How the sector performs on portfolio, engagement and resolutions*. https://influencemap.org/report/Climate-Funds-Are-They-Paris-Aligned-3eb83347267949847084306dae01c7b0

Macchiavello, E., & Siri, M. (2020). Sustainable finance and Fintech: Can technology contribute to achieving environmental goals? A preliminary assessment of 'Green FinTech'. *European Banking Institute Working Paper Series 2020*, 71. https://ssrn.com/abstract=3672989

McDaniels, J. (2020). *Implications of the COVID-19 pandemic for global sustainable finance*. https://unepinquiry.org/publication/implications-of-the-covid-19-pandemic-for-global-sustainable-finance/

OECD. (2019). *Blockchain technologies as a digital enabler for sustainable infrastructure*. https://www.oecd-ilibrary.org/docserver/0ec26947-en.pdf?expires=1621327878&id=id&accname=guest&checksum=EB2022D73B1D7B0CAC01E1C4D002397E

UNEP. (2016). *Fintech and sustainable development: Assessing the implications*. https://90bf1b73-3763-483f-abf5-1e464e9f1319.filesusr.com/ugd/3d4f2c_f57f7bc8bab34ecc8600cb19d3b52469.pdf

UN Secretary General's Task Force on Digital Financing of the Sustainable Development Goals. (2019a). *Harnessing the digitalization of finance for the sustainable development goals, White Paper, 28*. https://digitalfinancingtaskforce.org/wp-content/uploads/2019a/07/DFET-White-PaperFinal-08-17-afa.pdf

UN Secretary General's Task Force on Digital Financing of the Sustainable Development Goals. (2019b). *Harnessing digitalization in financing of the sustainable development goals', Progress Report*. https://digitalfinancingtaskforce.org/wp-content/uploads/2019b/09/Task-Force-CoChairInterim-Report

Postface

This book offers a comprehensive and critical review of the EU regulatory framework on sustainable finance, focussing on the role of information as a key driver of the transition towards sustainable investments and sustainable growth.

Lawmakers and securities regulators contribute to this complex transition developing and promoting transparency through disclosure of material ESG-related information as a key legal strategy.

The combined impact of the EU regulatory framework developed so far and constantly in progress is that of enabling all market participants (issuers, investors, financial institutions and firms) to identify and assess sustainability-related risks and opportunities and make sustainable investment choices.

A stronger regime of non-financial disclosure at the European level along with attentive supervision and targeted investor education initiatives facilitate this result.

A renewed understanding of the general materiality clause, shaped by the purpose and goals of the disclosure, provides a key component of this legal strategy, giving relevance to the information needs of the rising socially and environmentally responsible investment movement. This is the first building block of a legal strategy which must be pursued by lawmakers and securities regulators all over the world to meet the

N. Linciano et al. (eds.), *Information as a Driver of Sustainable Finance*, Palgrave Studies in Impact Finance,
https://doi.org/10.1007/978-3-030-93768-3

core goals of securities markets regulation (investor protection, market efficiency and integrity).

Convergent supervision is the second safeguard to ensure effective compliance with the regulatory framework and tackle risks of greenwashing and products mis-selling, which in turn are the key to ensure market integrity and allocative efficiency. In this respect, given the increasing volume of information and the need to ensure its comparability, quality and reliability, all opportunities offered by the technological innovation should be exploited: artificial intelligence, machine learning, distributed ledger technology may act as a digital enabler for sustainable infrastructure, reduce the costs of searching for information and improve the measurement and tracking of the sustainability of investments. In this perspective, the so-called GreenTech would benefit industry, regulators as well as investors.

Lastly, all stakeholders should be fully engaged in the path towards sustainability. Investors should be supported in developing knowledge and skills that are essential to make sound financial decisions without neglecting their preferences for sustainability. Manufacturers and distributors of financial products need to improve knowledge and skills to produce and distribute products that fit the preferences for sustainability of their target markets. Securities regulators and supervisors need to build new supervisory capacity, which means not only to have a good knowledge of the relevant regulations but also to understand how to embed the new sustainable requirements into the paths of the traditional supervision.

This book pictures the complexity of this overall fast evolving framework where several key players, including the securities regulators, should intensively cooperate, and where legal strategies aim at achieving a common final goal towards sustainable growth.

[DATE]

Chiara Mosca
Commissione Nazionale per le Società e la Borsa
Rome, Italy

Index

N. Linciano et al. (eds.), *Information as a Driver of Sustainable Finance*, Palgrave Studies in Impact Finance,
https://doi.org/10.1007/978-3-030-93768-3

FSC
www.fsc.org
MIX
Papier | Fördert gute Waldnutzung
FSC® C083411

Kolibri 360 GmbH
Ferdinand-Jühlke-Straße 7
99095 Erfurt, Deutschland
produktsicherheit@kolibri360.de